de-testing + de-grading schools

Studies in the Postmodern Theory of Education

Shirley R. Steinberg
General Editor

Vol. 451

The Counterpoints series is part of the Peter Lang Education list.
Every volume is peer reviewed and meets
the highest quality standards for content and production.

PETER LANG
New York • Washington, D.C./Baltimore • Bern
Frankfurt • Berlin • Brussels • Vienna • Oxford

de-testing + de-grading schools

AUTHENTIC ALTERNATIVES TO ACCOUNTABILITY AND STANDARDIZATION

EDITED BY JOE BOWER AND P.L. THOMAS

INTRODUCTION AND CHAPTER BY ALFIE KOHN

PETER LANG
New York • Washington, D.C./Baltimore • Bern
Frankfurt • Berlin • Brussels • Vienna • Oxford

Library of Congress Cataloging-in-Publication Data

De-testing and de-grading schools: authentic alternatives to accountability
and standardization / edited by Joe Bower, P.L. Thomas.
pages cm. — (Counterpoints: studies in the postmodern theory of education; vol. 451)
Includes bibliographical references.
1. Educational tests and measurements.
2. Educational accountability. 3. Educational change.
I. Bower, Joe, editor of compilation. II. Thomas, P. L. (Paul Lee), editor of compilation.
LB3051.D447 371.26—dc23 2013003336
ISBN 978-1-4331-2240-8 (hardcover)
ISBN 978-1-4331-2239-2 (paperback)
ISBN 978-1-4539-1081-8 (e-book)
ISSN 1058-1634

Bibliographic information published by **Die Deutsche Nationalbibliothek**.
Die Deutsche Nationalbibliothek lists this publication in the "Deutsche
Nationalbibliografie"; detailed bibliographic data is available
on the Internet at http://dnb.d-nb.de/.

The paper in this book meets the guidelines for permanence and durability
of the Committee on Production Guidelines for Book Longevity
of the Council of Library Resources.

29 Broadway, 18th floor, New York, NY 10006
www.peterlang.com

Printed in the United States of America

To Garfield High School (Seattle, WA) faculty
for their 2013 challenge to MAP testing

CONTENTS

Acknowledgments

This volume is a work of collaboration and intersections. As an edited work and collection of voices addressing de-testing and de-grading schools, the collaboration is obvious. The intersections are unique to this era of New Media since the volume began as a Twitter connection between Joe Bower and P. L. Thomas (volume editors) who shared in part a powerful influence by Alfie Kohn (Introduction and foundational chapter, "The Case Against Grades").

From Joe Bower:
Thanks to Paul Thomas for so graciously inviting me to be a part of this project; to Alfie Kohn for his guidance and patience with my emails and questions; to all the courageous contributors who work so hard to make school a better place by moving away from tests and grades; to my parents for their unconditional support; and to Tamara, Kayley, and Sawyer for putting up with my blogging and tweeting.

From P. L. Thomas:
Thanks to Joe Bower and Alfie Kohn for helping get this project rolling; to my students over the past three decades who have taught me many lessons about the negative impacts of grades and tests; to the education community on Twitter and throughout the world of blogs; and to Joe Kincheloe, in whose gracious wake I continue to work.

Notices:
Introduction copyrighted by Alfie Kohn. Used by permission.

Chapter One, "NCLB's Lost Decade for Educational Progress: What Can We Learn from This Policy Failure?," reprinted by permission from Guisbond, L., with Neill, M., & Schaeffer, B. (2012, January). NCLB's Lost Decade for Educational Progress: What Can We Learn from This Policy Failure? FairTest: National Center for Fair and Open Testing. Available at http://fairtest.org/NCLB-lost-decade-report-home.

Chapter Three, "Technocratic Groupthink Inflates the Testing Bubble," adapted from Living in Dialogue blog post (http://blogs.edweek.org/teachers/living-in-dialogue/2012/06/technocratic_groupthink_inflat.html), *Education Week*. Used by permission.

Chapter Seven, "Standardized Testing and Boredom at an Urban Middle School," adapted from an article originally published in *Perspectives on Urban Education* 9(1). Used by permission.

Chapter Ten, "The Case Against Grades," copyright © 2011 by Alfie Kohn. Reprinted by permission from Kohn, A. (2011, November). The case against grades. *Educational Leadership*. Available at http://www.alfiekohn.org/teaching/tcag.htm. This is a slightly expanded version of the published article.

Chapter Eleven, "Reduced to Numbers: From Concealing to Revealing Learning," author retains copyright.

Chapter Fifteen, "'Parents Just Want to Know the Grade': Or Do They?," adapted from article originally published in *Phi Delta Kappan* (September 2012). Used by permission.

Introduction

The Roots of Grades-and-Tests

Alfie Kohn

Most of the contributions to this book focus on problems with either grades or tests. In an article about college admissions published more than a decade ago, however, I suggested that we might as well talk about "grades-and-tests" (G&T) as a single hyphenated entity (Kohn, 2001). There are certainly differences between the two components, but the most striking research finding on the subject is that students' G&T primarily predicts their future G&T—and little else. It doesn't tell us much at all about their future creativity, curiosity, happiness, career success, or anything else of consequence.

In fact, the case for the fundamental similarity of grades and tests runs deeper than their limited predictive power. Both are "by their nature reductive," as P. L. Thomas, co-editor of this volume, observes in his chapter. I would add that both emerge from—and, in turn, contribute to—our predilection for three things: quantifying, controlling, and competing. All of these are defining characteristics of our educational system but also permeate our culture more generally.

To quantify is to talk about something in numerical terms. That's not a problem when a question lends itself to counting ("How large are elementary school classrooms as opposed to high school classrooms?") but becomes more troubling in the case of other inquiries ("How do we know if that teacher is any good?"). Just over a hundred years ago, Edmond G. A. Holmes, the chief inspector of elementary schools for Great Britain, remarked, "As we tend to value the results of education for their measurableness, so we tend to undervalue and at last ignore those results which are too intrinsically valuable to be measured" (as cited in Madaus & Clarke, 2001, p. 93).

Tests, or at least those that yield a score, are, like grades, based on the premise that learning can and should be quantified. Indeed, the pervasiveness of G&T suggests that the (reasonable) question "How should we assess . . . ?" has morphed into the (more problematic) question "How should we measure . . . ?"—as if assessment without numbers was either (a) so obviously inferior as to be undeserving of discussion or (b) simply impossible, because to assess *means* to measure (Kohn, 2012).

In his book *Trust in Numbers*, historian Theodore Porter (1995) points out that quantification has long exerted a particular attraction for Americans. "The systematic use of IQ tests to classify students, opinion polls to quantify the public mood, . . . even cost-benefit analyses to assess public works—all in the name of impersonal objectivity—are distinctive products of . . . American culture" (p. 147). I don't know whether this is more true in education than it is in other fields, but it seems particularly disturbing to assume that the process by which children come to make sense of ideas is always something we can count. And that assumption reveals itself not only through the ubiquity of G&T but also through more recent (and equally reductive) developments such as rubrics, which have the effect of smuggling in standardization through the back door (Wilson, 2006).

The enterprise of assessing and evaluating requires teachers to do two things: collect information about how students are doing and then share that information with the students and/or their parents. But tests aren't necessary to do the first and grades aren't necessary to do the second. A teacher who is paying attention—listening to students' conversations, following their projects, reading their writing—will never need to administer a test. (This assumes that students have a chance to converse, design projects, and write. If they're forced to spend their time listening and filling out worksheets, well, then there's not much authentic learning to *be* assessed.) In fact, the attentive teacher will acquire a broader and deeper understanding of how students are faring, and which of them need help with what, than he or she could with a test.

Tests are not only unnecessary but also unhelpful because they mostly tell us how many forgettable facts have been crammed into short-term memory, and how skilled students have become in the specialized art of test-taking. Steven Wolk (1998), an Illinois teacher, put it this way:

> In the real world of learning, tests and reports and worksheets aren't the most meaningful way to understand a person's growth, they're just *convenient* ways in a system of schooling that's based on mass production. . . . I assess my students by looking at their work, by talking with them, by making informal observations along the way. I don't need any means of appraisal outside of my own observations and the student's work, which is demonstration enough of their thinking, their growth, their knowledge, and their attitudes over time. (pp. 111–112)

Once the teacher has figured out the extent to which students' thinking is becoming more sophisticated and where gaps still exist, there's obviously no need to reduce the conclusion to a summary letter (B) or a number (84) or a label that functions just like a letter or number but allows us to pretend we're doing something different ("exceeds expectations"). Instead, a qualitative description or evaluation can be offered in narrative form—or, better yet, as part of a dialogue during a meeting with students or parents.

Why is G&T still so common if it's unnecessary and, as many of the chapters that follow in this book argue, downright harmful? Possible answers include tradition, the appeal of quantification (with its siren call of objectivity), a lack of familiarity with alternatives, and, as Wolk points out, simple convenience. But here's another explanation: Unlike more authentic ways of determining and then describing students' progress, G&T appeals to those who seek control. If I don't know how to work with my students to create a classroom and a curriculum that will pique their intellectual curiosity and persuade them to participate, I can simply coerce them into doing whatever I say—show up on time, sit down, and be quiet; write down what I say; read these pages or complete these exercises (at a pace I impose); do even more schoolwork at home—by warning them that noncompliance will result in their faring poorly on a test, which, in turn, will bring down their grade.

Extrinsic inducements, of which G&T is the classic example in a school setting, are devices whereby those with more power induce those with less to do something. G&T isn't needed for assessment, but it is nearly indispensable for compelling students to do what they (understandably) may have very little interest in doing. The same is true of standardized tests as a matter of public policy, particularly when rewards or punishments hinge on the results. This is how federal officials make state officials race to what they define as the top, how state officials make district administrators adopt a set of prescriptive curriculum standards, how administrators deprofessionalize teachers by compelling them to follow scripted lessons, and so on. (For readers who are already familiar with how high-stakes testing serves as a mechanism of control, what may be the new insight here is that the same is true of teacher-designed tests and quizzes, which are instruments by which teachers treat their students much as they complain about being treated themselves.)

As the engine of both school "reform" at the macro level and in-class assessment at the micro level, then, G&T creates spurious precision, flattening education into something that can be measured, and forces people to participate whether they like it or not. But it also has a third effect, which is to foster competition. Educational and psychological tests were invented to sort people—not just to rate but to rank. The original imperative wasn't to learn about test-takers in order to help them, but to determine who was better than whom and, practically speaking, which of them to select and which to leave behind.

Despite this history, it is possible to test in such a way that the results will not be used to pit students against one another for recognition or rewards—although testing remains problematic for other reasons. Similarly, grading needn't be done on a curve; the system can be set up so all students, at least in theory, may earn the top grade. (Grades would still function as extrinsic motivators, but at least there wouldn't be an artificial scarcity of A's.) Yet in practice G&T never seems to be too far removed from competition: Quantified results create an irresistible temptation to compare students. Even schools that prohibit teachers from grading on a curve may use grades to compute class rank, and the students themselves may feel compelled to keep asking one another, "Wad-ja-get?"

It's not a coincidence that defenders of G&T point to our competitive culture (recast as "the real world") to justify the practice. At the same time, those who are troubled by the effects of competition tend to be critical of G&T as well, and vice versa. Specifically, teachers who are committed to cooperative learning (as well as to democratic classrooms and the kind of thinking that can't be reduced to numbers) are also, in my experience, apt to steer clear of G&T whenever possible.

The distinguishing feature of that opposition is that G&T, and the underlying adherence to quantification, control, and competition, is understood as a problem in itself. We have to look beyond real but marginal objections to the way G&T has been implemented. The problem with testing isn't limited to what's *on* the test (or, even less important, whether the results are released in time to "do any good"). The problem with grading isn't limited to how many students get A's, or what role homework or class participation plays in determining the final grade, or whether it's possible to retake a test, or whether marks are posted online. Nor will replacing norm-referenced with criterion-referenced tests, or letter grades with rubrics, do the trick. The problem runs deeper, so our willingness to question and confront the status quo must follow suit.

References

Kohn, A. (2012, September 19). Schooling beyond measure. *Education Week*. Retrieved from www.alfiekohn.org/teaching/edweek/sbm.htm.

Kohn, A. (2001, March 9). Two cheers for an end to the SAT. *Chronicle of Higher Education*. Retrieved from www.alfiekohn.org/teaching/sat.htm

Madaus, G., & Clarke, M. (2001). The adverse impact of high-stakes testing on minority students. In G. Orfield & M. Kornhaber (Eds.), *Raising standards or raising barriers?: Inequality and high stakes testing in public education* (pp. 85–106). New York: Century Foundation.

Porter, T. M. (1995). *Trust in numbers: The pursuit of objectivity in science and public life*. Princeton, NJ: Princeton University Press.

Wilson, M. (2006). Rethinking rubrics in writing assessment. Portsmouth, NH: Heinemann.

Wolk, S. (1998). *A democratic classroom*. Portsmouth, NH: Heinemann.

PART ONE

Degrading Learning, Detesting Education

The Failure of High-Stakes Accountability in Education

ONE

NCLB's Lost Decade for Educational Progress

What Can We Learn from This Policy Failure?[1]

Lisa Guisbond with Monty Neill and Bob Schaeffer

Ten years have passed since President George W. Bush signed No Child Left Behind (NCLB), making it the educational law of the land. A review of a decade of evidence demonstrates that NCLB has failed badly both in terms of its own goals and more broadly. It has neither significantly increased academic performance nor significantly reduced achievement gaps, even as measured by standardized exams.

In fact, because of its misguided reliance on one-size-fits-all testing, labeling and sanctioning schools, it has undermined many education reform efforts. Many schools, particularly those serving low-income students, have become little more than test-preparation programs.

It is time to acknowledge this failure and adopt a more effective course for the federal role in education. Policymakers must abandon their faith-based embrace of test-and-punish strategies and, instead, pursue proven alternatives to guide and support the nation's neediest schools and students.

The data accumulated over ten years make three things clear:

1. NCLB has severely damaged educational quality and equity, with its narrowing and limiting effects falling most severely on the poor.

2. NCLB failed to significantly increase average academic performance and significantly narrow achievement gaps. And,

3. So-called "reforms," such as the Obama Administration's waivers and the Senate Education Committee's Elementary and Secondary Education Act (ESEA) reauthorization bill, fail to address many of NCLB's funda-

mental flaws and in some cases will intensify them. These proposals will extend a "lost decade for U.S. schools."

Despite a decade's worth of solid evidence documenting the failure of NCLB and similar high-stakes testing schemes, and despite mounting evidence from the U.S. and other nations about how to improve schools, policymakers cling to discredited models. This is particularly tragic for families who hoped their children's long wait for equal educational opportunity might be ending. It is also tragic for our public education system, whose reputation has been sullied by promises not kept and expensive intervention schemes that do more harm than good.

It is not too late to revisit the lessons of the past ten years and construct a federal law that provides support for equity and progress in all public schools. With that goal in mind, this report first provides an overview of the evidence on NCLB's track record. Second, it looks at recent efforts at NCLB "reform" and what past evidence says about their likely outcomes. Finally, it points to alternative strategies that could form the basis for a reauthorized federal law that would improve all schools, particularly those serving our most needy students.

Part I. The Record: NCLB's Promises Unmet

NCLB's ten-year report card offers little cause for celebration, whether you judge the law narrowly on its own terms or look more deeply at its impact.

- NCLB's own narrow gauges of progress reveal major shortcomings: growth on the National Assessment of Educational Progress (NAEP) has stalled, achievement gaps are stagnant, and predictions of widespread school "failure" are coming true.
- The curriculum has narrowed, test preparation has displaced broader schooling, cheating is rampant, there is too little help for schools in need, and NCLB has contributed to the growth of a pernicious school-to-prison pipeline.
- A narrow focus on testing and punitive accountability has caused policymakers to ignore the real educational consequences of child poverty, which has grown significantly in recent years.

Growth Stalled, Gaps Remain

Instead of helping to create circumstances in which schools can provide a rich, well-rounded curriculum and address the needs of individual students, the law has pressed schools to narrow curriculum, teach to the test, and resort to deceptive and unethical ways to boost test scores. It has done so by defining student learning and school quality in the narrow terms of standardized exam results.[2]

NCLB's chief yardsticks for measuring results are state standardized tests in math and reading administered annually in grades 3 through 8 and once in high school. The law designated NAEP tests as an independent yardstick. School leaders and teachers correctly feared that failure to meet state test targets could result in sanctions for their schools. With so much riding on the results, many schools turned to preparing students for these tests, ignoring other aspects of education.

Not surprisingly, scores on state-administered tests have shown greater growth than NAEP, on which scores have tended to stagnate. However, as benchmarks moved higher, stretching toward the goal of 100% proficiency, more and more schools in almost every state have fallen short. This is due in large part to the law's requirement that every one of multiple groups—race/ethnicity, low-income, English language learner and disabled—make "Adequate Yearly Progress" (AYP). In the 2010–2011 academic year, 48% of the nation's 100,000 schools failed to reach AYP benchmarks.

What about the backup measure? NAEP, too, is a standardized test, primarily multiple-choice with some short-answer questions. It has been particularly criticized for its flawed definition of "proficiency." Nevertheless, it is a technically sound standardized exam, generating consistent scale scores from year to year, allowing their use as an independent yardstick to track whether and when improvements have occurred.

The latest NAEP results (NCES, 2011a,b) confirm trends identified over the past decade (FairTest, 2009). Overall, growth on NAEP was more rapid before NCLB became law and flattened after it took effect. For example, 4th grade math scores jumped 11 points between 1996 and 2003, but increased only 6 points between 2003 and 2011 (see Fig. 1.1). Reading scores have barely moved in the post-NCLB era. Fourth grade scores increased just 3 points to 221 between 2003 and 2011, remaining level since 2007. In 8th grade reading, there was a meager 2-point increase, from 263 to 265, in that same period. Since the start of NCLB, gains have stagnated or slowed for almost every demographic group in both subjects and both grades.

As a result, gaps between groups remain large, despite the hope that NCLB's exposure of these gaps would motivate successful efforts to close them. In fact, gaps have remained mostly stagnant for most groups of students at both grade levels in both subjects. For example, in 8th grade math, the large gap between Whites and Blacks remained at 32 points from 2007 to 2009, closing by just one point in 2011. In 8th grade reading, Wisconsin is the only state that narrowed the gap between Whites and Blacks between 1998 and 2011, and only two states, Alabama and California, narrowed the gap between Whites and Hispanics.

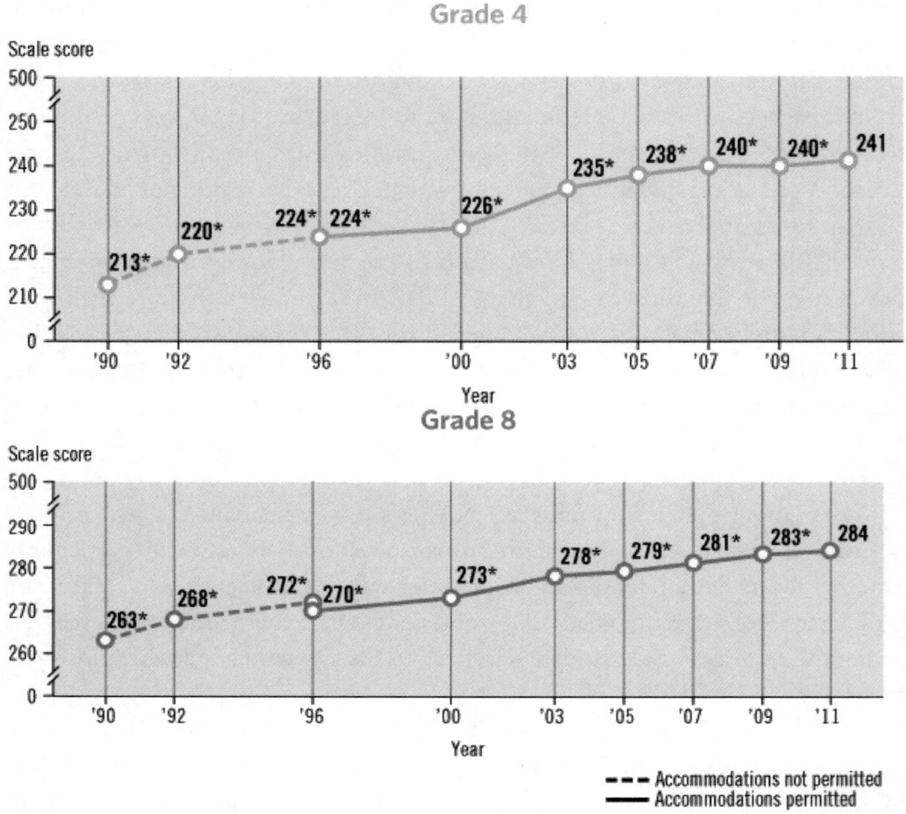

Figure 1.1. Trends in fourth- and eighth-grade NAEP math average scores (Source: NCES, 2011a)

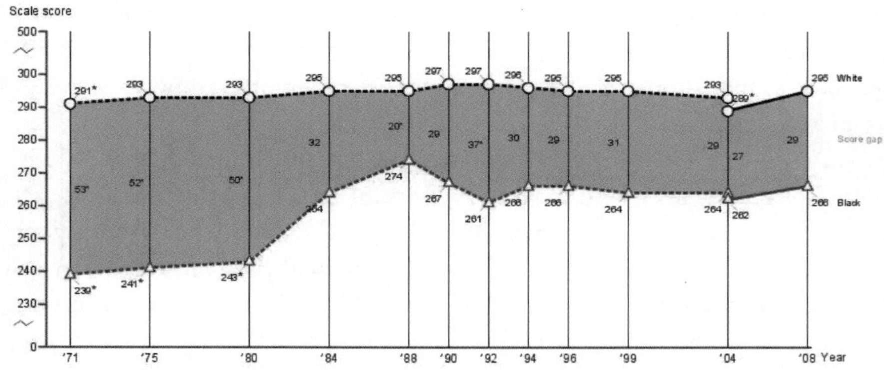

Figure 1.2. Long-term NAEP trends (Source: NCES, 2008)

Columbia University Professor of Sociology and Education Aaron Pallas (2011) looked at changes in the performance of White, Black and Hispanic students in every state on 4th and 8th grade reading and math between 2003 and 2011. He concluded that NAEP "provides no evidence that states can meet the laudable goal of convergence of student-subgroup performance at a significantly higher level of academic proficiency than is currently observed. No state over the past eight years has succeeded in doing this in the way that NCLB demands" (Pallas, 2011).

In fact, long-term NAEP trends show just one period in which achievement gaps narrowed dramatically (see Fig. 1.2). That era of strong progress toward education equity preceded not only NCLB but most state high-stakes testing polices.

In the Classroom: Overtesting, Curricular Narrowing, Teaching to the Test, Cheating and Other Forms of Corruption

NLCB demanded results in the form of test data, though the bottom-line results have fallen short. The law succeeded, however, at transforming many schools into highly focused, "data-driven" environments. Testing and test preparation have proliferated—the amount of time spent on testing in some schools has doubled. A study for Congress by the Government Accountability Office (GAO) estimated states would have to create more than 433 tests (at a cost of $1.9 billion to $5.3 billion between 2002 and 2008) to satisfy NCLB mandates (U.S. GAO, 2003). This has become just the tip of the iceberg of a massive increase in testing. It is not uncommon for 20 to 60 school days per year to be spent in test-preparation, on top of the days spent on testing itself, which are considerable. In Massachusetts, for example, there will be 33 state test sessions across all grades this year (DESE, 2011). While the benefits of this transformation are scant, the educational costs are extremely high.

One cost is the disruption of instructional time for students who need it most. The Wisconsin Association for Supervision and Curriculum Development attempted to quantify the learning time lost to testing in general and for students with special needs. They found Wisconsin teachers spent a per-district average of 976 hours administering tests. This was particularly damaging to special needs students: "Some schools reported that disadvantaged student populations experienced as many as 15 days—three weeks—of disrupted instructional services because the specialists were involved in test administration. Across a student's 12-year span in a district, that could result in as many as 36 weeks, or a full year, of disrupted services for the disadvantaged students who are at the greatest risk of not meeting NCLB objectives" (Zellmer et al., 2006).

Sometimes it takes an outright scandal for NCLB-induced learning losses to come to light. At Dallas, Texas's Field Elementary School, students were assigned grades in subjects they were not even taught in order to hide the school's exclusive focus on NCLB's tested subjects. This earned the school an "exemplary" rating from the state, but students were getting an education full of holes. A report from the Dallas Independent School District's Office of Professional Responsibility (OPR) included testimony from Field teachers, who were directed by the principal to set aside music, art, and science instruction. In an email explaining why one math/science teacher should focus on math, the principal wrote, "Since the kids are so low in math, [the teacher] has to stick with math. . . . This is a very high-stakes year and we cannot afford to have students' TAKS scores drop in third grade" (OPR, 2011). Teachers testified that they argued against these directives, but felt they would lose their jobs if they did not comply.

Then there is the cheating epidemic that has erupted across the nation. In Atlanta, where cheating was confirmed in 44 public schools, involving 178 teachers and principals, a Georgia Bureau of Investigation (GBI) report described a culture of "fear, intimidation and retaliation spread throughout the district" (GBI, 2011a). As 2011 came to a close, Georgia investigators released another report documenting widespread cheating on tests in Dougherty County, 200 miles south of Atlanta. They found evidence of cheating in each of the county's 11 schools and similar evidence of teachers coerced into correcting students' wrong answers. The report cited three main causes of the cheating. Reason number one: "Pressure to meet adequate yearly progress under the No Child Left Behind Act."

Such stories of corruption and cheating in the NCLB era are so common that they cannot be dismissed as the actions of a few individuals. Instead, they are a predictable, inevitable outcome of pressure to meet test score targets, regardless of circumstances. According to published reports, incidents of cheating in the past three years have been confirmed in 30 states and the District of Columbia, whose former chancellor Michelle Rhee has taken her "boost the scores" campaign national.

This cheating epidemic and other forms of corruption are classic examples of Campbell's law (1976), which states, "The more any quantitative social indicator is used for social decision-making, the more subject it will be to corruption pressures and the more apt it will be to distort and corrupt the social processes it is intended to monitor."

The National Research Council (NRC) of the National Academy of Sciences looked at the accumulated evidence on test-based policies, including the federal No Child Left Behind law, state graduation tests, and policies that give teachers bonuses if their students' scores go up (Hout & Elliott, 2011). The report concluded that test-based incentives like those in NCLB increase teaching to the test and produce an inflated and inaccurate picture of what students know. It also found that educators facing sanctions tend to focus on actions that improve test scores, such as teaching test-taking strategies or drilling students closest to meeting proficiency cutoffs, rather than improving learning.

There is copious evidence of NCLB's narrowing effects from a range of sources (see, for example, Au, 2007; McMurrer, 2007; NASBE, 2003; NCES, 2007). Common Core (2011) released preliminary results of a teacher survey in December 2011. It found that 66% of teachers said NCLB's focus on math and reading has meant reduced time for art, science, and social studies. Other reports have documented how many schools are cutting recess in order to expand test preparation time, even for young children.

"During the past decade, our public schools have focused—almost exclusively—on reading and math instruction" under No Child Left Behind, said Lynn Munson, president and executive director of Common Core. Though NCLB "clearly identifies our 'core curriculum' as reading, math, science, social studies, and even the arts," many subjects have been "abandoned," Munson explained. "As a result, we are denying our students the complete education they deserve and the law demands" (Common Core, 2011).

Most troubling is that the law has exacerbated inequities it promised to end. A report from the Council for Basic Education (Von Zastrow & Janc, 2004) found evidence that narrowing was most severe in schools with higher numbers of minority and low-income students.

Linda Perlstein explained what this looks like in her book *Tested: One American School Struggles to Make the Grade* (2007). Perlstein spent a year at Tyler Elementary, a low-income school in Anne Arundel (Maryland) County school district:

> That children from well-off families and children from poor ones have divergent school experiences is nothing new. What is significant is that the disparity continues in spite of (and in some ways because of) a movement designed to stop it. The practice of focusing on the tested subjects of reading and math at the expense of a well-rounded curriculum is far more prevalent where children are poor and minority. "You're not going to be a scientist if you can't read," a superintendent once told me in defense of a school's pared-down curriculum. Well, you can't be a scientist—one of the most common career goals of Tyler Heights' graduating fifth-graders—if you never learn science either. (p. 135)

NCLB's Role in Student Pushouts and the Growing School-to-Prison Pipeline

As bad as NCLB's narrowing and trivializing effects is the pushing out of low-scoring students to improve a school's test score bottom line. Sharon Nichols and David Berliner (2007) compiled substantial evidence of this in their book *Collateral Damage*, including in Birmingham, Alabama, where 500 students were dropped from high school before test time, and New York City, where a lawsuit exposed policies that pushed out thousands of low-scoring students. These practices, which disproportionately affect students of color and students with disabilities, are linked to the rapid growth of a "School-to-Prison Pipeline," which is driving more and more students into the criminal justice system. The swelling of the pipeline has more than one cause, to be sure, but a 2011 position paper produced by several civil rights and education groups explained the role played by the federal testing mandate. "NCLB had the effect of encouraging low-performing schools to meet benchmarks by narrowing curriculum and instruction and de-prioritizing the educational opportunities of many students. Indeed, No Child Left Behind's 'get-tough' approach to accountability has led to more students being left even further behind, thus feeding the dropout crisis and the School-to-Prison Pipeline" (Advancement Project et al., 2011, p. 1).

Too Much Blame, Too Little Support for Improvement

Some of NCLB's flaws might be forgiven if they had led to sustainable improvements for many schools and students in need. Instead, the law's flawed approach to accountability laid the foundation for an equally flawed and ineffective approach to providing options for parents and improving schools. A major piece of this was the provision allowing parents to transfer their children out of schools not making AYP into district schools that are. A December 2004 GAO report found fewer than 1% of the students eligible to transfer under the law did so in the 2003–2004 school year. A second NCLB remedy, the supplemental services provision, has funneled money to private tutoring businesses with no measurable positive effect on students. "NCLB's Supplemental Educational Services: Is This What Our Students Need?" reported that NCLB's supplemental education services were reaching just 233,000, or 11%, of the two million students eligible nationwide, frequently offering low-quality services that merely extend NCLB's "narrowed educational agenda into students' out-of-school hours" (Ascher, 2006, p. 136).

NCLB does not invest in building new schools in failing districts, nor does it make wealthy, higher performing districts open their doors to students from poor districts. Instead, it created a menu of restructuring options for schools that fail to make Adequate Yearly Progress for six consecutive years. Such schools are subject to one of the following: takeover of the school by the state; turning management of the school over to a private firm; shutting down and reopening as a charter school; or reconstitution of the school by replacing some or all administrators, staff, or faculty. A fifth option provided under the law endorses "any other major restructuring of a school's governance arrangement."

Researcher William J. Mathis looked at the record for these types of interventions in a 2009 brief. Overall, he found that there was not much of a track record for any of these approaches being used to restructure failing schools. When they were used, there was little evidence of success. For example, charter schools are rarely selected as a restructuring option, and, in any case, the record shows that "when controlling for demographic factors, charter schools show no advantage."

Mathis (2009, p. 17) concluded: "Given that these approaches are being proposed for the nation's most troubled schools, the solutions [currently set forth by NCLB] are likely to be woefully inadequate." What's more, states have no capacity to implement such sweeping restructuring remedies. While some NCLB proponents thought that the law would force states to reallocate or raise new funds to assist low-income, low-scoring schools, in general this has not been the case. Many schools remain seriously underfunded, and great funding inequalities exist both within districts and between districts in a state (FEA, 2011).

Focus on Testing Avoids Addressing the Consequences of Child Poverty

One reason why NCLB was doomed to fall short of its lofty goals has little to do with its flawed provisions or implementation. Right in the middle of the path to "100% proficiency" came the worst economic crisis since the Great Depression. According to a recent U.S. Census Bureau (2011) report, child poverty has risen to 22%, with 96 of the largest 100 school districts reporting growth in the number of poor children. Meanwhile, both school resources and the social supports children need to learn and succeed in school (housing, family and community stability, medical and dental care) are shrinking. Thus, schools must educate more and more children for whom the foundations of school success are crumbling.

The demands for equal outcomes in an unequal society would have been a dangerous illusion even without the fiscal crisis. To expect schools to counter the far-reaching impact of child poverty, to expect schools to not only keep these children from falling further behind but to accelerate their academic growth in order to close gaps in achievement, as NCLB does, is to deny reams of evidence of how poverty affects children's ability to learn, going back to the landmark Coleman report (1966). Coleman could not have foreseen the last few decades' staggering growth in income inequality and child poverty. But those constructing policy prescriptions during the past decade should have considered the many ways in which poverty influences a child's ability to learn (Rothstein, 2004; Berliner, 2009).

The book *Whither Opportunity?* edited by Professors Greg Duncan and Richard Murnane (2011) documents this rising income and educational inequality and the ways in which they are linked. In their *Chicago Tribune* op-ed, Duncan and Murnane (2011, October 6) explained:

> Growing economic inequality contributes in a multitude of ways to a widening gulf between the educational outcomes of rich and poor children. In the early 1970s, the gap between what parents in the top and bottom quintiles spent on enrichment activities such as music lessons, travel and summer camps was approximately $2,700 per year (in 2008 dollars). By 2005–2006, the difference had increased to $7,500. Between birth and age 6, children from high-income

> families spend an average of 1,300 more hours than children from low-income families in "novel" places—other than at home or school, or in the care of another parent or a day care facility. This matters, because when children are asked to read science and social studies texts in the upper elementary school grades, background knowledge is critical to comprehension and academic success.

Advocates who call attention to the influence of poverty on educational outcomes are accused of making excuses for schools' failure to close achievement gaps. The "No Excuses" proponents accuse these advocates of saying poor children cannot learn.

This charge is a red herring. A great deal can and should be done to improve schools. However, NCLB failed to consider the consequences of poverty and has been an excuse for not addressing them. Indeed, the educational "reforms" advanced in the law have predictably failed to improve schools or learning (Neill, Guisbond & Schaeffer, 2004).

On this tenth anniversary, there is ample evidence that NCLB's false premises—that high-stakes testing coupled with sanctions would improve outcomes, without having to address other educational issues or issues of poverty—have caused the law to fail. Clearly, it is past time for a major change of course. The problem is that most proposals on the table are more of a change in rhetoric than a change in substance.

Part II. Sour Wine with New Labels

Secretary Duncan has clearly heard a chorus of complaints about NCLB's ill effects—from superintendents, principals, teachers, parents, students, community activists and researchers. As a result, he often speaks of flawed or limited tests and narrowed curriculum, echoing these pervasive complaints. Yet his proposals and initiatives continue most of the worst aspects of NCLB and add new, equally unsupported and harmful ones.

- Neither the Administration's waivers nor a bill from the Senate Health, Education, Labor and Pensions (HELP) Committee propose to reduce the massive overuse of standardized testing that has followed in the wake of NCLB.

- Both proposals would abandon the destructive and unrealistic AYP provision, but both keep or even expand too much of what has not worked. Duncan's Race to the Top program and waiver proposals, in particular, show the Administration's failure to consider the evidence that explains why NCLB failed.

Role of Testing Grows, Unabated

Beyond its basic testing mandates, NCLB begot a seemingly endless proliferation of tests and ways to use them: standardized tests in more subjects, interim and benchmark tests. It spawned so-called "formative" tests, which are supposed to help improve instruction but mostly take more time away from it. NCLB also fed the growth of a hugely profitable testing industry, increasing its bottom line while student achievement on NAEP leveled off and achievement gaps stagnated. Sec. Duncan heard many calls for relief from the testing avalanche and at times seemed ready to answer them.

For example, the U.S. Department of Education claimed one motivation for the waiver program was that: "NCLB has put too much emphasis on a single standardized test on a single day. This is teachers' biggest complaint about the law. They feel pressure to prepare students for those tests, leading to an unintended narrowing of the curriculum and an emphasis on the basic skills measured by standardized tests" (USDOE, 2011).

Given that rationale, the obvious response of the Administration would be to relieve this pressure. In one sense, it appears to do this, by eliminating NCLB's detested Adequate Yearly Progress mechanism. However, the dominant role of testing remains firmly in place and will likely intensify in most states.

Under the waiver plan, states must continue annual testing in reading and math of all children in grades 3–8, and once in high school, but with new tests based on "college and career standards." The evidence thus far is that the new tests will largely resemble current tests—but be harder to pass. The waivers also require states to adopt "student growth" measures and make them a "significant factor" in teacher and principal evaluation. This has pushed states to adopt statistical techniques that research shows are grossly inaccurate. Even worse, it intensifies the focus on boosting test scores instead of ensuring the all-around education of the whole child. In other words, it perpetuates the false notion that you can fatten the pig by weighing it more frequently.

The Administration says that for subjects in which a state does not have tests, in order to measure "growth" there will need to be, if not more tests, then "measures that are comparable" within a district. This could push districts to buy or create dozens of new exams, at great expense and likely great damage to now-untested subjects. Charlotte-Mecklenburg, North Carolina, for example, allocated $1.9 million to create 52 new tests for teacher evaluation (Grundy & Sawyer, 2011; *The Herald Weekly*, 2011).

The waiver plan is particularly dangerous because most states (39 plus the District of Columbia) either have applied or say they will apply for a waiver (McNeil, 2011). Moreover, it appears unlikely that the Elementary and Secondary Education Act (ESEA), now labeled NCLB, will be reauthorized before the 2012 elections. The waiver plan could thus dramatically shape schooling in the U.S. and tighten the grip of testing (Foley & Neill, 2011).

Some states, such as California, have rejected the waivers, for reasons ranging from potentially harmful effects on education to the high costs states and communities will face to implement the waivers even as the fiscal crisis is forcing them to cut their budgets. Perhaps prior to submitting their waiver proposals, more states will decide it is a bad deal and put pressure on Sec. Duncan to waive the AYP requirements without the quid pro quo, or poison pill, of using student test scores to judge teachers and other unwarranted waiver schemes.

The Senate HELP bill also failed to scale back the central role of testing in a reauthorized ESEA. Its bill maintains all NCLB testing; functionally defines "achievement" as test scores; and uses scores as the near-sole basis for many educational decisions (Foley & Neill, 2011). But there are additional ways in which so-called reforms ignore existing evidence and threaten to exacerbate the damaging role of testing in school.

A New High-Stakes Tool: Linking Student Test Results to Teacher Evaluations

Despite multiple studies demonstrating that linking student test scores to teacher evaluations is unfair, inaccurate and not ready for prime time, Sec. Duncan decided to carry this controversial requirement forward from Race to the Top (RTTT)[3] to the NCLB waiver program. Once again, rather than solve the widely acknowledged problems of teaching to the test, narrowing the curriculum and perpetuating cheating and other types of corruption, this "innovation" will exacerbate them by making test results even more high stakes for teachers.

In November 2011, the Education Writers Association released a brief on teacher evaluation based on more than 40 studies and interviews with scholars (Sawchuk, 2011). The brief concludes that existing research does not support linking teacher evaluations to student test scores, for multiple reasons:

- Teachers are not the most important factor in student achievement, which is mostly a product of individual and family background characteristics.
- The politically popular value-added methods of measuring teachers are generally not reliable or stable. These measures may pick up some differences in teacher quality, but they can be influenced by a number of factors, including statistical controls and characteristics of schools and peers.
- Contrary to claims that student achievement can be greatly influenced by having highly effective teachers several years in a row, a teacher's effectiveness varies from year to year. The impact of an effective teacher seems to decrease with time, so the cumulative effects of having better teachers for several years in a row are not clear.

- In the United States, rewarding teachers whose students produce gains (sometimes termed "merit pay") has not been shown to improve student achievement.

University of California–Berkeley economist Jesse Rothstein (2011) offered the most succinct assessment of value-added methods, concluding they are "only slightly better than coin tosses" at measuring teacher quality.

Despite the evidence (see also Neill, 2011), because of pressure and incentives from RTTT and the NCLB waiver program, states have aggressively moved forward in planning and implementing teacher evaluation programs linked to student test results. At least 23 states and Washington, D.C., evaluate teachers in part by test scores, and 14 states allow districts to use data to dismiss teachers.

Tennessee was a winner in the RTTT competition, and swiftly implemented an evaluation system that bases half of a teacher's assessment on "student achievement," with 35% to come from growth measures based on student scores on state tests, and requires frequent evaluations by principals. The system caused such frustration and confusion that State Education Commissioner Kevin Huffman quickly called for modifications. One problem is that, as in most states, most teachers teach untested subjects, so there are no test scores to evaluate them. Tennessee "solved" this problem by allowing teachers to be evaluated with scores in a subject they do not teach. For example, a gym teacher could "choose" to be evaluated by students' scores in writing. Will Shelton, principal of Blackman Middle School in Tennessee, described the system: "I've never seen such nonsense. In the five years I've been principal here, I've never known so little about what's going on in my own building" (Winerip, 2011).

A similar New York plan has led to an unprecedented explosion of resistance from principals. As of late 2012, nearly 23% of New York State principals (1,058) had signed a protest statement objecting to "an unproven system that is wasteful of increasingly limited resources. More importantly, it will prove to be deeply demoralizing to educators and harmful to the children in our care. Our students are more than the sum of their test scores, and an overemphasis on test scores will not result in better learning" (N.Y. Principals, 2011).

Despite Lack of Success, Turnaround Strategies Stay on the Program

The Administration's waiver program and the Senate HELP bill eliminate AYP and instead create much more limited categories of schools requiring intervention. States granted waivers must focus turnaround efforts only on the lowest performing 5%, so-called "priority" schools. Another 10% ("focus" schools) would have interventions targeted at their lowest performing student groups, which could include the transfer and tutoring options that were unsuccessful in NCLB.

The Senate HELP bill similarly identifies the lowest 5% of schools, based on test scores and (for high schools) graduation rates, for interventions.

This nod to states' limited capacity was a small dose of reality. But the waiver program's turnaround alternatives hew closely to NCLB, which included reopening as a charter school, replacing most or all of the staff, turning governance over to an outside entity, or "any other major restructuring."

The waiver options include:

- A turnaround model that would replace the principal and rehire no more than half the school staff.
- A restart model in which the school is converted or closed and reopened under an education management organization, which could be a charter.
- The school could simply be closed and its students enrolled in other schools.
- A transformation model, "which address four areas critical to transforming persistently lowest-achieving schools. These areas include: developing teacher and principal leader effectiveness, implementing comprehensive instructional reform strategies, extending learning time and creating community connections, and providing operating flexibility and sustained support."

The latter model is a softer version of RTTT requirements and at least offers the possibility of taking genuine improvement steps. No federal funding is provided, however, for these costly school improvement approaches.

The Senate HELP bill's turnaround strategies are also similar, though the list includes an additional three options that could allow more flexibility (Foley & Neill, 2011). In addition, HELP requires districts to develop specific improvement actions for all its "5%" schools, based on a review of the institution and rooted in such things as professional development and collaboration that have proven to lead to genuine school improvement when done well. To some meaningful degree, here the Committee responded to recommendations from education and other organizations (FEA, 2011b).

Without evidence, administration officials turn to anecdote, highlighting "model" schools to show that their prescriptions have worked in the past. Education scholar Diane Ravitch, however, looked closely and found no miracles (Ravitch, 2011b). For example, the president traveled to Florida in March to join Gov. Jeb Bush in praising Miami Central High for its transformation, after more than half the staff had been fired. Ravitch found that this "miracle school" remains one of Florida's lowest performers and narrowly evaded closure.

While miracle turnarounds are rare, evidence-based strategies for improving school performance have shown success. This report's next section briefly summarizes alternative approaches to NCLB reform that include these real-world approaches.

Part III. Real Reform Is Possible but Would Mean Setting a New, Evidence-Based Direction

NCLB's authors tried to give the law a level of gravitas by calling for "scientifically based research" (SBR) to guide educational practice. The law defined SBR as "research that involves the application of rigorous, systematic, and objective procedures to obtain reliable and valid knowledge relevant to education activities and programs." Many, including FairTest (Neill, Guisbond & Schaeffer, 2004), argued from the start that existing research on high-stakes testing, turnaround strategies and other aspects of the law should have led NCLB's authors in very different directions. Unfortunately, these pleas had little effect.

Fortunately, there are models of successful practice here and abroad that could form the basis for a revised ESEA, which would support meaningful, sustainable educational reforms.

Linda Darling-Hammond (2010), Diane Ravitch (2011a), Tony Wagner (2011) and many others have observed that top-ranked Finland, for example, dramatically reformed its schools over 20 years to achieve a high degree of equity and quality. It did so by pursuing policies that diametrically oppose those in NCLB. Finland has no high-stakes testing to rank students or schools and does not evaluate teachers based on student test scores. There are no mass firings of teachers, closures of "failing schools" or turnaround experts brought in to shake things up. There are no scripted curricula or frequent benchmark assessments leading up to a big test.

Instead, Finland focuses on ensuring equitable educational resources across the board—even providing more to schools serving students with the greatest needs. It has developed a strong, unionized teaching force that works together to improve schooling, and a sound, comprehensive curriculum. Finland invests heavily in teacher preparation and development and then gives their well-prepared and supported teachers tremendous autonomy and respect. Other top-ranked nations such as Singapore and Hong Kong pursue similar approaches, and their students reap the benefits (Darling-Hammond, 2010).

It is more difficult to find successful comprehensive models in the U.S., in part because of a profoundly unequal society and in part because all public schools are ruled by NCLB's rigid requirements. But on a smaller scale, there are schools that demonstrate better methods of assessment, such as the use of multiple measures of student learning (FairTest, 2010a). For example, the New York Performance Standards Consortium (n.d.) high schools have taken advantage of a variance from the New York Board of Regents to limit standardized testing to just the state English Language Arts (ELA) exam. The schools fulfill state and federal accountability requirements using the ELA test along with their own math and language arts tasks and other performance-based assessments.

This flexibility has allowed these schools, whose demographic makeup is roughly comparable to New York City public schools, to create a system with extremely high expectations and outcomes. According to the Consortium's web site (n.d.), "The [performance] tasks require students to demonstrate accomplishment in analytic thinking, reading comprehension, research writing skills, the application of mathematical computation and problem-solving skills, computer technology, the utilization of the scientific method in undertaking science research, appreciation of and performance skills in the arts, service learning and school to career skills." Dropout rates are very low (9.9% compared to the citywide rate of 19.3%) and college acceptance rates high (91% compared with the citywide rate of 62.6%). It is easy to imagine how much more they and others could accomplish within the context of a federal law that actively encouraged and supported such an approach.

Alternative Proposals for NCLB Reform

More than 150 national education, civil rights, religious, disability, civic, labor and other groups have now signed the Joint Organizational Statement on NCLB. The statement enumerated problems with NCLB and recommended reforms, including a move away from the "overwhelming reliance on standardized tests to using multiple indicators of student achievement in addition to these tests."

Out of that initiative came the Forum on Educational Accountability (FEA), which in 2011 laid out a proposal for NCLB reauthorization (FEA, 2011b). The plan would ensure that schools have the capacity to help all children achieve success while outlining a reasonable federal role in educational policy instead of top-down mandates that are too often overly prescriptive and fail to help schools reach desired educational and societal goals. FEA's proposals cover four main areas: overhauling assessment, restructuring accountability, developing school capacity to serve all students well, and addressing the unmet human and social needs faced by many children. Each is rooted in solid evidence—research and practical experience from the U.S. and other nations.

Relying on work done in Massachusetts, FairTest (2010b) has promoted a three-part assessment and evaluation program each state could implement. The plan includes gathering and evaluating classroom- and school-based evidence of student learning each year for each school; administering low-stakes standardized statewide tests in reading and math to each student every few years; and using "school quality reviews" which involve teams of experts conducting a careful review of each school every 4–6 years to ascertain how well it is meeting the full range of student needs. Together, this evidence would provide a far richer picture of student learning and school progress than can standardized tests alone. At the same time, it would avoid the damaging consequences of NCLB such as teaching to the test and narrowing the curriculum.

Proposals from other groups, such as the Broader, Bolder Agenda for Education (2008), share fundamental characteristics with FairTest and FEA reform proposals that sharply distinguish them from NCLB and its offspring. These alternatives explicitly acknowledge the need for dramatic and fundamental, not incremental, changes to the law. They recognize that the purpose of assessment is to help teachers improve instruction and to strengthen schools, not to label and punish them. They support methods to identify needed improvements more precisely in order to better target school efforts and outside assistance. They recognize the need for educational and other public policies to address health, social, emotional and other basic needs. They call for the education of the whole child and for gathering a range of data relevant to that goal, including survey data on school climate.

NCLB was promoted as an example of a bipartisan consensus on education policy. Over time, it has become clear that the details of the law, if not some of its stated goals, were fundamentally flawed. This report has summarized the evidence of what went wrong as well as the nation's ongoing education policy challenge and how to confront it in order to meaningfully improve public education. After a decade of stagnation, it is now time to use this evidence to craft and pass a new federal education law that will help and not harm our schoolchildren.

Endnotes

1 Reprinted by permission from Guisbond, L., with Neill, M., & Schaeffer, B. (2012, January). NCLB's Lost Decade for Educational Progress: What Can We Learn from this Policy Failure? FairTest: National Center for Fair and Open Testing. Available at http://fairtest.org/NCLB-lost-decade-report-home

2 FairTest's report, Failing Our Children (Neill, Guisbond & Schaeffer, 2004), explains the myriad ways high-stakes testing damages the quality of education and undermines individual opportunities. In doing so, it explained why NCLB was going to leave many children behind.

3 RTTT offered competitive grants to states that agreed to comply with the Administration's favored policies, such as charter expansion, national standards and teacher evaluations linked to student test scores.

References

Advancement Project, et al. (2011, March). Available online at http://www.scribd.com/doc/51122359/Advancement-Project-School-to-Prison-Pipeline-Position-Paper

Ascher, C. (2006). NCLB's supplemental educational services: Is this what our students need? *Phi Delta Kappan, 88*(2), 136–141.

Au, W. (2007). High-stakes testing and curricular control: A qualitative metasynthesis. *Educational Researcher, 36*(5), 258–267.

Berliner, D. (2009). Poverty and potential: Out of school factors and school success. Available online at http://nepc.colorado.edu/publication/poverty-and-potential

Broader, Bolder Approach to Education. (2008). Available online at http://www.boldapproach.org/index.php?id=51

Campbell, D. T. (1976). *Assessing the impact of planned social change*. The Public Affairs Center, Dartmouth College.

Coleman, J., et al. (1966). *Equality of educational opportunity*. Government Printing Office.

Common Core/Farkas Duffet Research Group. (2011). Learning less: Public school teachers describe a narrowing curriculum. Highlights from a survey by Common Core and the Farkas Duffett Research Group. Available online at http://www.commoncore.org

Darling-Hammond, L. (2010). *The flat world and education: How America's commitment to equity will determine our future*. New York: Teachers College Press, Teachers College, Columbia University.

Duncan, G., & Murnane, R. (2011). *Whither opportunity? Rising inequality, schools and children's life chances*. New York: Russell Sage.

Duncan, G., & Murnane, R. (2011, October 6). Economic inequality: The real cause of the urban school problem. *The Chicago Tribune*. Available online at http://articles.chicagotribune.com/2011-10-06/news/ct-perspec-1006-urban-20111006_1_poorchildren-graduation-rate-gap

FairTest. (2010a). Multiple Measures: A definition and examples from the U.S. and other nations. Available online at http://www.fairtest.org/fact-sheet-multiple-measures-definition-and-exampl

FairTest. (2010b). A better way to evaluate schools. Available online at http://www.fairtest.org/fact-sheet-better-way-evaluate-schools-pdf

FairTest. (2009). NAEP Results Produce More Evidence of NCLB's Failure. Available online at http://www.fairtest.org/naep-results-produce-more-evidence-nclbs-failure

Foley, B., & Neill, M. (2011). Is AYP really dead? A look at waivers and ESEA reauthorization. Available online at http://fairtest.org/slide-show-basic-facts-aboutduncans-ayp-waiver-pl

Forum on Educational Accountability (FEA). (2011a). All children deserve the opportunity to learn. Available online at http://www.edaccountability.org/reports.html

Forum on Educational Accountability (FEA). (2011b). FEA recommendations for improving ESEA/NCLB. Available online at http://www.edaccountability.org/Legislative.html

Georgia Bureau of Investigation (GBI). (2011). Special investigation into CRCT cheating at APS. Available online at http://www.ajc.com/news/volume-1-of-special-1000798.html

Georgia Bureau of Investigation (GBI). (2011b). Dougherty County investigative report, vols. 1–2. Available online at http://www.gaosa.org/

Grundy, P., & Sawyer, C. (2011, May 15). The push behind a new flurry of testing. *The News Observer*. Available online at http://www.newsobserver.com/2011/05/15/1196022/the-push-behind-a-new-fl urry-of.html

The Herald Weekly. (2011). School district to collaborate with state on test development. Available online at http://www.huntersvilleherald.com/education/2011/12/22/school-district-to-collaborate-with-state-on-test-development/

Hout, M., & Elliott, S. (eds.). (2011). Incentives and test-based accountability in education. Committee on Incentives and Test-Based Accountability in Public Education; National Research Council. Available online at http://www.nap.edu/catalog.php?record_id=12521

Joint Organizational Statement on No Child Left Behind Act. (2004). Available online at http://www.edaccountability.org

Massachusetts Department of Elementary and Secondary Education (DESE). (2011). Massachusetts Comprehensive Assessment System Schedule for MCAS and MEPA Testing. Available online at http://www.doe.mass.edu/mcas/cal.html

Mathis, W.J. (2009). NCLB's ultimate restructuring alternatives: Do they improve the quality of education? Boulder and Tempe: Education and the Public Interest Research Center & Education Policy Research Unit. Available online at http://nepc.colorado.edu/publication/nclb-ultimate-restructuring

McMurrer, J. (2007). NCLB year 5: Choices, changes, and challenges: Curriculum and instruction in the NCLB era, Center on Education Policy. Available online at http://www.cep-dc.org/displayDocument.cfm?DocumentID=312

McNeil, M. (2011). Majority of states say they'll seek waivers under NCLB. *Education Week*. Available online at http://www.edweek.org/ew/articles/2011/10/13/08waivers.h31.html?r=953279945

National Association of State Boards of Education (NASBE). (2003). The complete curriculum. Ensuring a place for the arts and foreign languages in America's schools. The report of the NASBE study group on the lost curriculum. Alexandria, VA.

National Center for Educational Statistics (NCES). (2011a). The nation's report card: Mathematics 2011. Available online at http://nces.ed.gov/nationsreportcard/

National Center for Educational Statistics (NCES). (2011b). The nation's report card: Reading 2011. Available online at http://nces.ed.gov/nationsreportcard/

National Center for Education Statistics (NCES). (2008). The nation's report card: Long-term trend 2008. Available online at http://nces.ed.gov/pubsearch/pubsinfo.asp?pubid=2009479

National Center for Education Statistics (NCES). (2007). Changes in instructional hours in four subjects by public school teachers of grades 1 through 4 (issue brief). Available online at http://nces.ed.gov/pubsearch/pubsinfo.asp?pubid=2007305

Neill, M. (2011). Student test scores: An inaccurate way to judge teachers. Available online at http://fairtest.org/sites/default/files/Student_Test_Scores_dangerous_way_to_evaluate_teachers_-_4-24-11.pdf

Neill, M., Guisbond, L., & Schaeffer, R. (2004). Failing our children: How "No Child Left Behind" undermines quality and equity in education; An accountability model that supports school improvement. Available online at http://fairtest.org/node/1778

New York Performance Standards Consortium. (n.d.). Available online at http://www.performanceassessment.org

New York Principals. (2011). An open letter of concern regarding New York State's APPR legislation for the evaluation of teachers and principals. Available online at http://www.newyorkprincipals.org/

Nichols, S., & Berliner, D. (2007). *Collateral damage: How high-stakes testing corrupts America's schools*. Cambridge, MA: Harvard Education Press.

Office of Professional Responsibility. (2011). Report of investigation. Available online at http://www.scribd.com/doc/73055392/DISD-Field-Elementary-investigation

Pallas, A. (2011). The nation's report card and NCLB: Friends or foes? National Education Policy Center. Available online at http://nepc.colorado.edu/blog/nation%E2%80%99s-report-card-and-nclb-friends-or-foes

Perlstein, L. (2007). *Tested: One American school struggles to make the grade*. New York: Henry Holt.

Ravitch, D. (2011a, October 11). What can we learn from Finland? Available online at http://blogs.edweek.org/edweek/Bridging-Differences/2011/10/what_can_we_learn_from_finland.html

Ravitch, D. (2011b, December 6). Do you believe in miracles? Available online at http://blogs.edweek.org/edweek/Bridging-Differences/2011/12/do_you_believe_in_miracles.html

Rothstein, J. (2011). Review of "Learning about teaching: Initial findings from the measures of Effective Teaching Project." Boulder, CO: National Education Policy Center. Available online at http://nepc.colorado.edu/thinktank/review-learningabout-teaching

Rothstein, R. (2004). Class and schools: Using social, economic, and educational reform to close the black–white achievement gap. Washington, D.C.: Economic Policy Institute. Available online at http://www.epi.org/publication/books_class_and_schools/

Rothstein, R., et al. (2008). Narrowing the achievement gap for low-income children: A 19-year life cycle approach. Washington, D.C.: Economic Policy Institute. Available online at http://www.epi.org/publication/narrowing_the_achievement_gap_for_low-income_children/

Sawchuk, S. (2011, November 12). What studies say about teacher effectiveness. Education Writers Association. Available online at http://www.ewa.org/site/PageServer?pagename=research_teacher_effectiveness

U.S. Census Bureau. (2011). Income, poverty and health insurance coverage in the United States: 2010. Available online at http://www.census.gov/newsroom/releases/archives/income_wealth/cb11-157.html

U.S. Department of Education (DOE). (2011). The new consensus on middle-grades reform. (Remarks of U.S. Secretary of Education Arne Duncan to the Association for Middle Level Education (AMLE) Annual Conference.) Available online at http://www.ed.gov/news/speeches/new-consensus-middle-grades-reform

U.S. Government Accountability Office (GAO). (2003). Characteristics of tests will influence expenses; Information sharing may help states realize efficiencies. Available online at www.gao.gov/new.items/d03389.pdf

U.S. Government Accountability Office (GAO). (2004). Education needs to provide additional technical assistance and conduct implementation studies for school choice provision. Available online at http://www.gao.gov/products/GAO-05-7

Von Zastrow, C., & Janc, H. (2004). Academic atrophy: The condition of liberal arts in America's public schools. Council for Basic Education. Available online at http://www.ecs.org/html/Document.asp?chouseid=5058

Wagner, T. (2011). *The Finland phenomenon: Inside the world's most surprising school system.* Available online at http://www.2mminutes.com/products/pc/viewPrd.asp?idProduct=22&idcategory=24

Winerip, M. (2011). In Tennessee, following the rules for evaluations off a cliff. *The New York Times.* Available online at http://www.nytimes.com/2011/11/07/education/tennessees-rules-on-teacher-evaluations-bring-frustration.html?pagewanted=all

Zellmer, M., et al. (2006). What are NCLB's instructional costs? *Educational Leadership, 64*(3), 43–46. Association for Supervision and Curriculum Development (ASCD). Available online at http://www.ascd.org/publications/educational-leadership/nov06/vol64/num03/What-Are-NCLB%27s-Instructional-Costs%C2%A2.aspx

||TWO

High-Stakes Testing Assessment

The Deus ex Machina of Quality in Education

Fernando F. Padró

This is not an argument against high-stakes testing assessment as designed or practiced, mainly because many of the concerns and limitations discussed in the literature from the time of the conception of the No Child Left Behind Act are still valid. James Popham's (1999) arguments as to why standardized tests do not measure educational quality have been supported by Daniel Koretz's (2008) findings and echoed in the recommendations found in the National Academy of Sciences' (Board on Testing and Assessment; National Research Council, 2009) open letter to the U.S. Department of Education on their view of the Race to the Top Fund. The reader is best served to look at these documents directly because this author does not have more to add in this regard beyond the observation that in higher education, the challenges addressed in Koretz's book and the National Academy of Sciences letter reflect many of the challenges faced by the people involved in *Measuring Up 2008* (National Center for Public Policy and Higher Education, 2008), an attempt to generate a state report card for higher education when it comes to evaluating learning.

This discussion focuses on the notion of quality and those strands that contextualize the expectation of what quality is. As someone who has been looking at notions of quality in education from the lens of industry and then accreditation for many years, it has become apparent to me that the issues surrounding student learning and performance are not primarily of a technical nature; they are political because policy steering and regulatory compliance define and ensure that the aims of policy steering come to life. The best way to describe the misdirection of arguments in regards to educational aims and results seems to be best grounded in Pierre Bourdieu's (Bourdieu & Passeron, 2000) notions of *social capital* and *habitus*:

> The specific productivity of [pedagogic work] is objectively measured by the degree to which it produces its essential effect of inculcation, i.e. its effect of reproduction . . . [and] is capable of durably generating practices conforming with the principles of the inculcated arbitrary. (p. 33)

Why? Because of what Luhmann (1996) calls *interpenetration:* the capacity of different systems within the social structure are able to generate a reciprocal relationship that allows each system to "enable each other by introducing their own already-constituted complexity into each other" (p. 213), that is, educational systems are not recognized as part of the solution and their body of knowledge is not considered adequate to meet the requirements of policymakers.

Assessment is used as a proxy for quality. The question, then, is a simple one, but it generates many others. What is quality? What or who sets the expectation(s) that must be met? What is the justification for the expectations? How is quality achieved?

Those of us in education typically look at these issues from the perspectives of the accountability and assessment movements that have been around for quite some time. Many relegate the quality movement as a fad that has come and gone, but one must argue that this is the language of the industry quality movement that is influencing the discussion of learning and what is being learned, performance at educational institutions, definitions of successful teaching, and expected organizational behavior from educational institutions and educators. Evidence of this influence is seen in the legislative and regulatory language defining education in the United States and what is being proffered by international agencies interested in advancing education. While much of the discussion over the past number of years has centered on primary and secondary education, proof of similar issues in higher education can be seen in the United States, for example, in the different incantations of the Higher Education Reauthorization Act (HERA) of 1965 from 1992 onward and from an international perspective the communiqué from the 2009 UNESCO World Conference on Higher Education.

The creation of the National Advisory Committee on Institutional Quality and Improvement (NACIQI) within HERA formed a committee whose recommendations use the ideas and language from the quality model from industry to shape expectations and processes of determining the success, merit, and worth of colleges and universities. The Council on Higher Education Accreditation (CHEA) has been acting as a foil to minimize the intrusion of the industrial model into higher education, as can be seen in the changing language and review parameters in many of the accrediting bodies involved in accrediting institutions or programs. As illustrated in the most recent report at the time of this writing,

NACIQI (2012) considers a diverse range of topics and seeks input from different stakeholder groups: "from federal and state actors, from accreditors, from beneficiaries of quality in higher education, and from accredited institutions, including perspectives from experts in education, policy, business, government, and beyond" (p. 1). Nevertheless, as Judith Eaton, CHEA president, wrote in a letter responding to NACIQI dated March 16, 2012:

> We share NACIQI's view that accreditation should continue to be the primary vehicle for assuring and improving quality in higher education. At the same time, we believe that some recommendations in the report may lead to a federal standardization of expectations of academic quality. The report suggests a more active role by the U.S. Department of Education (USDE) in the practices of accreditation. These recommendations would move accreditation away from its focus on institutional academic leadership through peer review that has been integral to its success and effectiveness and would alter the relationship between institutions and accreditors in counterproductive ways. (open letter accessible from http://www.chea.org/Government/FedUpdate/CHEA_FU25.html)

Recommendations from NACIQI in many ways reflect views espoused today in the international arena. The 2009 World Conference on Higher Education (UNESCO, 2009) talks about the 21st-century world as one being a knowledge society. Within this culture,

> [t]he training offered by institutions of higher education should both respond to and anticipate societal needs. This includes promoting research for the development and use of new technologies and ensuring the provision of technical and vocational training, entrepreneurship education and programmes for lifelong learning. (p. 4)

Access, equality, and quality are linked into a critical triad in this communiqué:

> Quality assurance is a vital function in contemporary higher education and must involve all stakeholders. Quality requires both establishing quality assurance systems and patterns of evaluation as well as promoting a quality culture within institutions. (p. 4)

In sum, UNESCO member states are asked to "put in place and strengthen appropriate quality assurance systems and regulatory frameworks with the involvement of all stakeholders" (p. 8). Mihailescu (2004) probably represented this view best when he wrote:

> A predominant model of the university as an institution in service to society will focus on parameters and standards evaluating the efficient use of public funds, while with a prevailing entrepreneurial model, parameters and standards evaluat-

ing post-graduation results, including scientific innovation, technologies, and a highly qualified workforce will stand out. (p. 51)

From here forward the discussion reflects how assessment and quality are used as proxies for each other. The discussion comes more from a higher education viewpoint than a P–12 one, but one reason for this is that higher education is facing many of the same issues and pressures; therefore, the concerns at the macro level are more similar than dissimilar. In other words, it is another way of looking at those external influences affecting education and all aspects of educational activity from early childhood until the brink of formally entering the workforce. While the focus is not always on testing and assessment, the discussion is always about testing and assessment because that is the stock in trade within the quality model that is strongly affecting education.

The Quality Model's Definition of Quality

Quality as a concept is different from the constructs of quality assurance, quality control, and quality management in that its definition is what drives these constructs.

Quality Assurance

At its simplest, Juran (1999) suggests that the prime purpose of quality assurance is to serve the needs of "those who are not directly responsible for conducting operations but who have a need to know—to be informed as to the state of affairs and, hopefully, to be assured that all is well" (p. 2.13). Frazer (1992, as cited in Mishra, 2006, p. 82) suggests that quality assurance can be defined as four components:

1. everyone in the enterprise has a responsibility for enhancing the quality of the product or service;
2. everyone in the enterprise has a responsibility for *maintaining* the quality of the product or service;
3. everyone in the enterprise understands, uses and feels *ownership* of the systems which are in place for maintaining and enhancing quality; and
4. management regularly checks the validity of the system for checking quality.

Quality Control

In contrast, "[q]uality control has as its primary purpose to maintain control" (Juran & Godfrey, 1999, p. 4.3). The main purpose is to maintain stability through feedback loops. This regime imposes an external set of standards that internal goals, which are the bases of processes, must meet in terms of output and throughput (compliance) deliverables. With improvements, new controls have to be put in place to prevent the performance level from deteriorating (Juran & Godfrey, 1999).

Quality Management

Deming (1994) noted that transformation is required in government, industry, and education to move out of the present style of management, that a mere patchwork of the "same-old" approaches to leadership and management is not enough. Deming (1994, 2000/1986) even goes on to say that best effort is not enough; support of management is not sufficient unless a new philosophy geared toward significant improvement is in place to transform the enterprise.

According to Marquardt (1999),

> Quality Management is the umbrella framework for managing the quality of a product. The philosophy, managing procedures, and technology should provide an operational system [where the different components of an enterprise] can work together to meet increasingly stringent customer requirements. (p. 27.2)

Quality management is a means of generating both quality assurance and quality control function processes by building an institutional climate to support and document quality improvement. Worth noting is Feigenbaum's (1983) observation that the multifunctional aspect of quality management enhances the organization's ability to effectively interact with the market that helps define quality.

Quality

The American Society for Quality's (ASQ) Quality Glossary (http://asq.org/glossary/q.html) states that *quality* is a subjective term meaning different things to different people, but there are technical elements to it. Harvey and Green (1993) said it best: "It means different things to different people, indeed the same person may adopt different conceptualizations at different moments" (p. 10). It is precisely that problem of subjectivity that is the real gravamen of the differences of opinion regarding testing and assessment in general as a proxy for determining learning and, by extension, demonstrating quality. Table 2.1 (p. 33) identifies how five influential gurus from the quality world define quality as a general concept. It is because of the success these individuals helped create in the United States and Japan after World War II that their ideas gained credence outside of the world of industry. Japan's postwar economic success and the desire to revive the flagging U.S. economy during the 1970s and 1980s allowed the quality movement with its focus on production processes and product quality using statistical tools and analysis methods to expand "into all areas of business, including the service and public sectors" (Smith, 2011, p. 45). Although these seminal thinkers in the field at times disagree whether quality is absolute, relative, a process, or a culture onto itself (Mishra, 2006), Table 2.1 shares the foundations for defining quality that are driving the education reform movement critics and supporters:

- systems focused,
- customer and stakeholder focused,
- top management driven with strong employee participation steeped into the organizational culture,
- process based,
- data driven (with a strong preference for statistical analysis and the concepts of validity and reliability) evaluation processes based on standards/regulatory compliance to provide that external view and contextualization of the organization,
- enhancing agility to meet the changing external environment,
- concerned about continuous improvement and organizational learning (the fostering of innovation), and
- value enhancing.

For all of the technical discussion about what quality is, there is no way around the reality that "[q]uality is essentially a function of human psychology" (Judd, 1994, p. 135). Quality, therefore, also refers to the ongoing process of building and sustaining relationships by assessing and anticipating demands to fulfill stated and implied needs (Winder, 1993). It is all about generating *satisfaction* (trying to maximize it) rather than settling for what Simon (1976) called *satisficing*, seeking the alternative available at the moment that seems the best. "It raises the question of 'whose quality?'" (Harvey & Green, 1993, p. 10). Internal and external stakeholders "[e]ach have a different perspective on quality. This is not a different perspective on the same thing but different perspectives on different things with the same label" (p. 10).

Ironically, while the focus of the quality model is to satisfy market and policy demands through regulatory and standards compliance, the communications exchange is lessening for educators. There are fewer alternatives available because this body of knowledge is not given legitimacy of place in the discussion, reducing education's ability to interpenetrate the policy steering mechanisms currently in place. Instead, the exchange between educators, policymakers, and stakeholders is akin to the stance taken by the Borg in the *Star Trek* television series oeuvre: "You will be assimilated. Resistance is futile."

Phillip B. Crosby (2005, p. 61)	Quality means conformance, not elegance. There is no such thing as a quality problem. There is no such thing as the economics of quality; it is always cheaper to do the job right the first time. The only performance standard is zero defects. The only performance measurement is the cost of quality (COQ).
W. E. Deming	Quality can be defined only in terms of the agent (2000, p. 168). Quality begins with intent, which is fixed by management (p. 5). The customer is the most important part of the production line (2000, p. 174). Quality must be measured by the interaction between three participants . . . : (1) the product itself; (2) the user and how he uses the product . . . ; (3) instructions for use, training of customer and training of repairman (2000, p. 176). Performance and style, whatever these words mean in the minds of customers, must show constant improvement. Zero defects is not sufficient (1994, p. 12). If you can't describe what you're doing as a process, you don't know what you're doing (Quality Progress Staff, 2010, p. 17).
Armand V. Feigenbaum (1983, p. 7)	Quality is a customer determination, not an engineer's determination, not a marketing determination. It is based upon the customer's actual experience with the product or service, measured against his or her ***requirements***—stated or unstated, conscious or merely sensed, technically operational or entirely subjective—and always representing a moving target in a competitive market . . . . The purpose of most quality measurements is to determine and evaluate the *degree* or *level* to which the product or service approaches this total composite (p. 7).
Kaoru Ishikawa	The features of the quality system should be clarified at all levels of the organization and communicated in such a way that the people have confidence in these features (Watson, 2004, p. 54). Ishikawa created customer focus within the quality movement, and today, this is the fundamental starting point of quality (p. 55). He also said all work must include corrective and preventive action to uncover and resolve problems downstream from the customer engagement point, making it the most cost-effective way to operate (Quality Progress Staff, 2010, p. 19).
Joseph M. Juran	1. "Quality means those features of products which meet customer needs and thereby provide customer satisfaction..." 2. "'Quality' means freedom from deficiencies--freedom from errors that require doing work over again (rework) or that result in field failures, customer dissatisfaction, customer claims, and so on" (1999, pp. 2.1-2.2). The short phrase 'fitness for use' applies.

Table 2.1. Five Definitions of Quality

Deming's System of Profound Knowledge and the Limitations in Defining Quality for Education

Deming and Feigenbaum were two of the quality gurus advocating a philosophical approach to quality. In his last book, Deming (1994) came up with what he called the System of Profound Knowledge (SPK), which was a clarification of his earlier 14 points (Deming, 2000/1986) for the transformation of management into a philosophy of interdependency and cooperation to accomplish the aims of an organization. His approach was to look at organizational aims in terms of the operational and political considerations that have to be considered as they pursue quality in an environment of uncertainty.

Deming's philosophy is centered on core management processes, giving priority to high quality of products and services, continuous monitoring, everyone's involvement in continuous improvement process, process management practices, and defining standards with sufficient precision so that customers are able to understand what the firm is producing (Padró, 2009; Phelps, Parayitam, & Olson, 2007). Deming's management method arises from efforts on the part of organizational leaders to move toward the simultaneous creation of a cooperative and learning organization based on process-management practices that support customer satisfaction, sustained employee fulfillment, and the continued improvement of processes, products, and services (Anderson, Rungtusanatham, & Schroeder, 1994).

Deming's SPK provides a structure to follow in looking at how the quality movement expectations present challenges to traditional educational thought and language. Deming's SPK has four dimensions to it; however, the discussion is based on Padró's (2009) expansion to six dimensions to reflect different issues within two of Deming's original four dimensions. By looking at the six separate elements, it is possible to consider issues through the lens of what Argyris and Schön (1974) referred to as the intellectual problems related to integrating thought with action.

1. Appreciation for a system

According to Deming (1994), it is important for an organization and its employees to have an aim, interdependence, and integration. The aim of any system emerges from the needs and expectations of consumers of the services and must be clear to everyone in the system.

Using systems thinking is a way of identifying, linking, analyzing, and managing complex interactions between different entities. Events seem to involve sociocultural developments rather than individual decisions and actions (Bertalanffy, 1969). From a meta-level view, there is an obligation for all components of the system to contribute its best to it, not to maximize its own benefit (Deming, 1994). This argument suggests that it is the greater good that must prevail, that the big picture is more important than individual element success. The questions that come to the fore are (1) What does "the system" include? and (2) What is the degree of its self-determination and ability to continue making its own actions?

In education, the notion of a system has various practical implications. For example, one straightforward implication is the challenge to the paradigm of segmented educational levels by the idea of lifelong learning and the transitions to support a more seamless school-to-career pipeline metaphor. Braided within the paradigm shift is a more nuanced implication, namely, the control of choice—not in terms of access or equity, but in determination of educational opportunity to pursue and the type of knowledge a person must have. Both examples touch on the role of stakeholders: who controls the systems and who makes what choices.

The concept of stakeholders is not new to educators because of the covenant that has always existed between the school and the community at large. Different (at least in the United States) is the trend to shift away from local oversight to state and national supervision of schools, colleges, and universities. Recognizing segmentation of stakeholders between internal and external interests is not new, either. What is changing is the dynamic of who shapes the goals for education and educational attainment. Globalization is not as much the culprit per se rather than the *nexus* (or will it be *nexum*?) between education and economic well-being. The issue is whether education is replaced by training and teaching by instruction. If one uses Machlup's (1962) basic typology of knowledge—practical knowledge (useful in work, decisions, and actions), intellectual knowledge (satisfying intellectual curiosity), small-talk/pastime knowledge (non-intellectual curiosity and for entertainment), spiritual knowledge, and unwanted knowledge (of no interest to the person that is usually acquired by accident and somehow retained)—what is at play is the locus of control: the individual or the community as the representative of workforce and general well-being interests. Who determines what is practical knowledge and unwanted knowledge? What is the value and the shaping of intellectual knowledge? Tightening of resources, increased demand, and heightened expectation for economic benefit places the balance of power on community decision-makers rather than on individuals or educators.

2. *Knowledge about variation*

The focus is on variation and statistical process control. Deming (1994) points out that variation happens, either by common or special causes. What is important is to understand what causes the variation. Rather than place blame, the focus should be on improvement. Do not necessarily eliminate variation when it is important to maintain it.

From the get-go, variation is a major problem for educators, and this is not in reference to qualitative researchers. Anyone holding a constructivist mindset will have a problem (Padró, 2009) because not everyone learns the same thing the same way or understands the information to the same extent. Interestingly enough, many educational reform efforts are based on constructivism (Null, 2004; Winchester, 2007), but this generates a tension because the two models can be antagonistic in how the learning happens. The difficulty rests in that, as Whitehead (1929) suggests, learning reflects the dynamic process of *concrescence*, in which abstractions become concrete categories with a verifiable scheme as its objective. Success is based on how circumstances to which the scheme is valid are applicable. "The ultimate test is always, widespread, recurrent experience; and the more general the rationalistic scheme, the more important is this final appeal" (Whitehead, 1929, p. 25). The value of experiences does not consist "of externally presented material, but of interaction of native activities with the environment which progressively modifies both the activities and the environment" (Dewey, 1944, p. 79).

As seen above, the quality model is all about process control as a means of ensuring the success through attaining similar results, something the above paragraph suggests is difficult to do. There is a limit to how much process control limits variation because there are so many exogenous variables affecting learning in any setting, especially in the classroom, not to mention the strong possibility of students achieving different scores when given the same test over a period of time. Moreover, when measuring success, indicators beyond academic performance have to be considered such as the principles of good practice for assessing student learning suggests (Carr & Harris, 2001; Astin et al., 1993). Consequently, the indicators and measures used to determine successful learning do not do a good job of it when limited to only reporting high-stakes testing results. Applying Birnbaum's (1988) description of an anarchic institution to this discussion, organizational learning is challenged by problematic goals, unclear technology, and fluid participation as related to vision and control of institutional decision making.

Two concerns that crop up in determining the benefits of assessment are the problem of determining value-added benefits of learning and how to assure that assessment strategies do not discourage *deep learning* for the more immediate rewards of *surface learning* by allowing a focus on instructional interaction that motivates students to learn (cf. Entwistle & Smith, 2002). Astin's (1991) *I-E-O model* talks about the importance of recognizing that learning and growth only happen over an extended period of time, meaning that one cannot learn much from a onetime administration of tests. Furthermore, Astin agrees with Deming's view that it is important to understand why some learners learn more than others, a point that seems lost in the equation for what has become a formative process in summative clothing.

Conflict between the quality model and educators who do not drink from the education reform cup rests on the view that quality can be predicted (Deming, 1994). Educators embracing the ideas from quality seem to be returning to John Franklin Bobbit's view that a primary goal of curriculum design is to eliminate waste *qua* teaching people what they would never use (Shepard, 2000), leading back to the question of who determines what is useful and what is not. Process management and monitoring through the use of high-stakes tests or other instruments with similar goals often do not recognize or can account for external non-school variables that demonstrably impact learning. Education reformers impose identical expectations for improvement on schools with low test score results without making any effort to determine those particular factors affecting results (Koretz, 2008). Assessment advocates point out research that shows assessment has positive short-, middle-, and long-term effects on learning (Black & William, 2010/1998; Earl, Volante, & Katz, 2011). Effectively, the expectations versus realities of assessment of student learning have the actual outcome of making high-stakes testing nothing more than a deus ex machina.

3. *Theory of knowledge*

To Deming (1994), this dimension is about listening to the expertise found within the different units of the organization. It is about organizational learning for these voices, and their sources of data become the lenses through which performance can be understood. These are the lenses that help analyze the collected data and help predict the future. "This is analogous to Argyris and Schön's (1974) Theory of Action" (Padró, 2009, p. 12) in that it is about generating the organizational view that helps explain, predict, or control the organization's behavior.

The point here is a simple one: The educational body of knowledge (focus, research emphasis, methodology, applicability) is deemed inferior to other disciplines, the updated version of the Dewey versus Thorndike debate at the beginning of the 20th century. From the time of *A Nation at Risk* in 1983, educators are to blame for many of the shortcomings in learning and attainment plus their interests are not deemed sufficiently aligned to social expectations. Educators no longer speak the same language as the other sectors of society as already mentioned at the beginning of this work. The *habitus* Bourdieu (1993) talks about does not represent the dispositions in the field of production that the dominant elements of the society have and demand. The disconnect or miscommunication comes from not being heard and allowed to influence the decision-making process. Truly ironic is that Deming's (1994) SPK is adamant about listening to the body of knowledge, but education seems the exception to this rule.

4. *Psychology*

"At the core of understanding quality and its processes, from Deming's perspective, is personal transformation through meaning. System change is predicated on a person's ability and willingness to change," explains Padró (2009, p. 12).

Deming's approach is similar to Goleman's (1998) view of emotional intelligence because of the focus on personal motivation and the willingness to tackle change in light of the enterprise's need to reinvent itself. An institution's management should focus on the plusses brought forth by individual differences and focus on approaches that motivate individuals and generate reciprocal trust. When reciprocity is not there, motivation takes a turn from striving to one of survival, with action concerned with defensive, *minimaxing* strategies (cf. Padró, in press).

5. Interaction and interdependence

While interconnectedness is part of the SPK as envisioned by Deming, Scholtes (1999) proposed *interaction* and *interdependence* as a fifth element to the SPK because of the need to understand and influence the interactions and independence "among and between the system, variability, learning and human behavior" (p. 707) as a means to understand complex change. "Scholtes differentiates these two concepts from Deming's psychology dimension based on the ability of individuals to make the necessary connections among the internal units" (Padró, 2009, p. 12). Educators are willing to work hard to increase learning and improve personal performance. However, most are not formally prepared to interact with the public at the various arenas necessary (although the ELCC/ISSLC standards for educational leaders are an attempt to remedy this lack). More challenging is the ability to maintain interdependence rather than become dependent because of fiscal constraints demarcating the boundaries of participation in feedback and decision-making when it comes to the teaching of learners.

6. Public policy/Policy steering

Much of the literature on quality is based on regulatory compliance and standards formation as a means of reducing variability. This emphasis somewhat ignores the impact public policy through policy steering has on institutional performance. There is a taxonomic difference between policy and regulatory compliance based on purpose and impact on organizations. Lowi (1972) suggested a difference between the purpose of prediction and the normative business of choosing a selected outcome because of the need to recognize one or more behavior patterns at the model rather than incremental levels as the basis of policy choice. Government acts as the analyst of what is needed, transforming taste to judgment because proper distance is established (Arendt, 1982) based on one of the four options available to policymakers (Lowi & Ginsberg, 1990) to decide appropriateness (quality) (Padró, 2009).

Mechanisms of policy steering define quality and are the basis for regulations. This is in contrast to regulatory compliance and standards formation whose purpose is to reduce and eliminate variation of these expectations. The challenges for educators are not the standards or the rules. It is not whether these reflect minimal demands or exemplary performance. Rather, they are what these represent and the process that controls them (cf. Padró, 2010). Because of this the status of the professional associations representing all aspects of the education sector is important. So too is the ability for education and educators to impact and control at least how standards are generated and put in place. Watch out for external agents not directly associated with education who prepare standards that they expect should be applied to educational endeavors based on their notions of performance, even if the language and ideas do not represent accepted standards of or best practices.

Concluding Comments: Irony at Work

Large-scale external testing existed in the United States from the late 19th century onward (Hamilton & Koretz, 2002). The assessment movement as such seriously began in the 1920s and has continued to the present day in spite of ups and downs regarding the appropriateness and usefulness of many of the instruments. Higher education began its interest in assessment in the 1980s, although its interest is of a less immediate concern. The accountability movement in education is of a more recent vintage, springing forth out of the passing of the Elementary and Secondary Education Act of 1965 (Lessinger, 1973). Initially the concerns of accountability may have been fiscal in nature, but it soon took on the task of assessing student achievement and typing it to social goals, its current primary emphasis since No Child Left Behind took effect. Both of these movements have been around long enough to become part of the education framework, and each is based on the language and theories that make up the framework of education as a field of study.

Quality as a model (or movement) began in the United States, moved as it were to Japan, and came back to rekindle the country's economic engine. While many of the seminal ideas have been around since the 1920s and 1930s, the quality movement did not generate interest outside education until about 1980 and in education by the end of that decade. However, the quality model as discussed in this narrative never has fully taken hold, having been labeled a fad whose time came and went.

The purpose and language of these three realms do cross over and tend to complement each other more often than not. What is noteworthy is that the interests and concepts of both the accountability and assessment movements can be embedded within quality thinking as exemplified by the Malcolm Baldrige National Quality Award Criteria. Just as important, however, are that all three movements share the limitations that critics of the accountability and assessment movements point out to whoever is willing to listen. What is different is that quality is based on a different theoretical framework, that of business and industry: the scientific management and human resources schools of thought. As a result, quality brings a different mindset based on engineering (putting together) and production rather than on bio-psycho-social considerations. Fundamentally, the difference is on what learning entails based on the competing views of whether learning processes can ensure that all learners acquire and master information to the same extent or if there are variables that cannot be controlled that distinguish learner accomplishment.

The discussion presented above contextualizes how this distinction affects education from the lens of quality and its philosophical framework to highlight where convergence occurs and where differences are significant. Throughout the narrative, the underlying theme is that quality model sets the tone of the language that defines and bounds educational practice in today's world. Clearly, this means that the quality movement is not a fad; instead, it has become one of the dominant voices in policy steering. Its impact on education may be indirect, more so because the accountability and assessment efforts within school can easily be subsumed into the larger framework and thus its voice is confused with these other two. Nevertheless, its views and techniques are dominant, and what is being felt is that subsummation that is eroding education's autonomy.

There is an irony at play here: In the quest for "meaningful data," the emphasis is on creating an incorruptible, objective testing process that yields such results. The challenge here is what is meaningful for one pair of eyes is warrantless in the sight of another pair. However, it is important to understand how the perspectives from the field of quality affect the discussion of assessment practices and procedures in schools. Its approach is well-intentioned, but it is incomplete because the emphasis on results takes away from those external environmental issues beyond scholastic control that influence the results. No process is fully controlled to mitigate or eliminate these variables, and therein rests the challenge in embracing this model to the fullest. On the other hand, there are lessons that can be learned. Knowing how the quality movement affects policy steering is important for proponents of the assessment movement as well as those interested in challenging that view and who prefer a learning environment that is not centered on tests and grades. In both instances, it is all about engendering a "meaningful" understanding of learning in the classroom and, through translation, later in the students' personal and professional lives.

References

Anderson, J. C., Rungtusanatham, M., & Schroeder, R. G. (1994). A theory of quality management underlying the Deming management method. *Academy of Management Review, 19*(3), 472–509.

Arendt, H. (1982). *Lectures on Kant's political philosophy*. R. Beiner (Ed.). Chicago: University of Chicago Press.

Argyris, C., & Schön, D. A. (1974). *Theory in practice: Increasing professional effectiveness*. San Francisco: Jossey-Bass.

Astin, A. W. (1991). *Assessment for excellence: The philosophy and practice of assessment and evaluation in higher education*. New York: American Council on Education and Macmillan.

Astin, A. W., et al. (1993). Principles of good practice for assessing student learning. *Leadership Abstracts, 6*(4), 2-3.

Bertalanffy, L. v. (1969). *General system theory: Foundations, development, applications.* (Revised ed.). New York: George Braziller.
Birnbaum, R. (1988). *How colleges work: The cybernetics of academic organization and leadership.* San Francisco: Jossey-Bass.
Black, P., & William, D. (2010/1998). Inside the black box—raising standards through classroom assessments. *Phi Delta Kappan, 92*(1), 81–90.
Board on Testing and Assessment; National Research Council. (2009). *Letter report to the U.S. Department of Education on the Race to the Top Fund.* Washington, DC: National Academies Press. Retrieved from http://www.nap.edu/catalog.php?record_id=12780
Bourdieu, P. (1993). *The field of cultural production: Essays on art and literature.* New York: Columbia University Press.
Bourdieu, P., & Passeron, J-C. (2000). *Reproduction in education, society and culture* (2nd ed.). London: Sage.
Carr, J. F., & Harris, D. E. (2001). *Succeeding with standards: Linking curriculum, assessment, and action planning.* Alexandria, VA: Association for Supervision & Curriculum Development.
Crosby, P. B. (2005, May). Crosby's 14 steps to improvement. *Quality Progress, 38*(5), 60–64.
Deming, W. E. (1994). *The new economics for industry, government, education* (2nd ed.). Cambridge, MA: MIT Press.
Deming, W. E. (2000/1986). *Out of the crisis.* Cambridge, MA: MIT Press.
Dewey, J. (1944). *Democracy and education.* New York: Free Press.
Earl, L., Volante, L., & Katz, S. (2011). Unleashing the promise of assessment for learning. *Education Canada, 51*(3), 17–20.
Entwistle, N., & Smith, C. (2002). Personal understanding and target understanding: Mapping influences on the outcomes of learning. *The British Psychological Society, 72,* 321–342.
Feigenbaum, A. V. (1983). *Total quality control* (3rd ed.). New York: McGraw-Hill.
Frazer, M. (1992). Quality assurance in higher education. In A. Craft (Ed.), *Quality assurance in higher education* (pp. 9–25). London: Falmer.
Goleman, D. (1998). *Working with emotional intelligence.* New York: Bantam.
Hamilton, L. S., & Koretz, D. M. (2002). Tests and their use in test-based accountability systems. In L. Hamilton, B. Stecher, & S. P. Klein (Eds.), *Making sense of test-based accountability in education* (pp. 13–49). Santa Monica, CA: Rand Corporation.
Harvey, L., & Green, D. (1993). Assessment and evaluation in higher education. *Assessment & Evaluation in Higher Education, 18*(1), 9–34.
Judd, D. K. (1994). The psychology of quality. In *American Society for Quality Control 48th Annual Quality Congress Proceedings* (pp. 134–139). Milwaukee, WI: American Society for Quality.
Juran, J. M. (1999). How to think about quality. In J. M. Juran, A. B. Godfrey, R. E. Hoogstoel, & E. G. Schilling (Eds.), *Juran's quality handbook* (5th ed., pp. 2.1–2.18). New York: McGraw-Hill.
Juran, J. M., & Godfrey, A. B. (1999). The quality control process. In J. M. Juran, A. B. Godfrey, R. E. Hoogstoel, & E. G. Schilling (Eds.), *Juran's quality handbook* (5th ed., pp. 4.1–4.29). New York: McGraw-Hill.
Koretz, D. (2008). *Measuring up: What educational testing really tells us.* Cambridge, MA: Harvard University Press.
Lessinger, L. M. (1973). Accountability and humanism: A productive educational complementarity. In C. D. Sabine (Ed.), *Accountability: Systems planning in education* (pp. 3–18). Homewood, IL: ETC Publications.
Lowi, T. J. (1972). Four systems of policy, politics, and choice. *Public Administration Review, 32*(4), 298–310.
Lowi, T. J., & Ginsberg, B. (1990). *American government: Freedom and power.* New York: W.W. Norton.
Luhmann, N. (1996). *Social systems* (J. Bednarz Jr. & D. Baecker, Trans.). Stanford, CA: Stanford University Press.

Machlup, F. (1962). *The production and distribution of knowledge in the United States.* Princeton, NJ: Princeton University Press.

Marquardt, D. W. (1999). Process industries. In J. M. Juran, A. B. Godfrey, R. E. Hoogstoel, & E. G. Schilling (Eds.), *Juran's quality handbook* (5th ed., pp. 27.1–27.37). New York: McGraw-Hill.

Mihailescu, I. (2004). The quality assessment and accreditation of higher education in Central and Eastern Europe. In L. Vlasceanu & L. C. Barrows (Eds.), *Indicators for institutional and programme accreditation in Higher/Tertiary Education* (pp. 39–54). Bucharest: UNESCO-CEPES. Retrieved from www.aqa.ac.at/download.69.unesco-cepes-indicators.pdf

Mishra, S. (2006). *Quality assurance in higher education: An introduction.* Bangalore, India: National Assessment and Accreditation Council.

National Advisory Committee on Institutional Quality and Improvement. (2012, April). *Report to the U.S. Secretary of Education: Higher Education Act Reauthorization, accreditation policy recommendations.* Retrieved from http://www2.ed.gov/about/bdscomm/list/naciqi-dir/2012-spring/teleconference-2012/naciqi-final-report.pdf

National Center for Public Policy and Higher Education. (2008). *Measuring up 2008: The national report card on higher education.* San José, CA: Author. Retrieved from http://measuringup2008.highereducation.org/print/NCPPHEMUNationalRpt.pdf

National Commission on Excellence in Education. (1983). *A nation at risk: The imperative for educational reform.* Washington, DC: United States Government Printing.

Null, J. W. (2004). Is constructivism traditional? Historical and practical perspectives on a popular advocacy. *Educational Forum, 68*(2), 180–188.

Padró, F. F. (2009). Deming's System of Profound Knowledge and its application to higher education institutions. *Journal for Quality Progress, 32*(1), 10–14.

Padró, F. F. (2010, June). *How quality assurance impacts the use of student evaluation of faculty: A changing paradigm that shifts the locus of control from the faculty to students and external stakeholders.* Paper presented at the Annual Conference of the American Association of University Professors, Washington, DC.

Padró, F. F. (in press). Accreditation and institutional learning: A stochastic proposition because it is often pursued from the perspective of a minimaxing regime. In M. Shaw & C. S. Nair (Eds.), *External quality audits: Has it improved quality assurance in universities?* Cambridge, UK: Chandos.

Phelps, L. D., Parayitam, S., & Olson, B. J. (2007). Edwards Deming, Mary P. Follett, and Frederick W. Taylor: Reconciliation of differences in organizational and strategic leadership. *Academy of Strategic Management Journal, 6*, 1–14.

Popham, W. J. (1999). Why standardized tests don't measure educational quality. *Educational Leadership, 56*(6), 8–15.

Quality Progress Staff. (2010, November). Guru guide: Six thought leaders who changed the quality world forever. *Quality Progress, 43*(11), 14–21.

Scholtes, P. R. (1999). The new competencies of leadership. *Total Quality Management, 10*, 704–710.

Secretary of Education's Commission on the Future of Higher Education. (2006, September). *A test of leadership—Charting the future of U.S. higher education: A report of the Commission appointed by Secretary of Education Margaret Spellings.* Jessup, MD: Education Publishing Center.

Shepard, L. A. (2000). The role of assessment in a learning culture. *Educational Researcher, 29*(7), 4–14.

Simon, H. A. (1976). *Administrative behavior: A study of decision-making processes in administrative organization.* New York: Free Press.

Smith, J. (2011, October). The lasting legacy of the modern quality giants. *Quality, 50*(10), 40–47.

United Nations Educational, Scientific and Cultural Organization (UNESCO). (2009, July 8). *2009 World Conference on Higher Education: The new dynamics of higher education and research for societal change and development.* Paris: Author. Retrieved from http://www.unesco.org/fileadmin/

MULTIMEDIA/HQ/ED/ED/pdf/WCHE_2009/FINAL%20COMMUNIQUE%20 WCHE%202009.pdf

Whitehead, A. N. (1929). *Process and reality*. New York: Macmillan.

Winchester, I. (2007). Editorial: Construction and education. *Journal of Educational Thought, 41*(1), 1–5.

Winder, R. E. (1993). *Fulfilling quality's five dimensions*. Paper presented at the 47th Annual Quality Congress, Transactions, in Atlanta, GA.

||THREE

Technocratic Groupthink Inflates the Testing Bubble*

Anthony Cody

As I was taking a look at the latest report from the National Council on Teacher Quality (NCTQ), a disturbing thought came to mind. The NCTQ has prepared a report that criticizes schools of education for failing to jump on the "obsessed with data" bandwagon. You can just feel the irritation in the words of NCTQ president Kate Walsh when she says:

> A lot of schools of education continue to become quite oppositional to the notion of standardized tests, even though they have very much become a reality in K–12 schools. The ideological resistance is critical.

This reminds me of a phenomenon called *groupthink*. I think what we are experiencing in education is actually a virulent and coercive strain of groupthink, and it is doing our students a great deal of harm. As I wrote in 2011 (Cody, 2011, April 28), the value of test data has been inflated way beyond their true worth, in a manner similar to real estate prices during the bubble of the past decade. Once this bubble is launched, many people begin to depend on it for their livelihoods (Resmovitz, 2012).

There's been a flood of tax-exempt corporate money for advocacy, think-tank research, and lobbying to direct public education policy into a public-private partnership under corporate, not public, control. This has in turn produced a network of consultants, paid strategists, leveraged public administrators and legislators, media pundits, and academic grantees. They now owe their positions and their livelihoods to an insular and self-affirming pattern of groupthink.

* Adapted from Living in Dialogue blog post (http://blogs.edweek.org/teachers/living-in-dialogue/2012/06/technocratic_groupthink_inflat.html), *Education Week*. Used by permission.

First, let's take a look at how groupthink is defined. The Psychologists for Social Responsibility offer this description, drawing on the work of Irving Janis, who has documented eight symptoms of groupthink:

1. Illusion of invulnerability—Creates excessive optimism that encourages taking extreme risks.
2. Collective rationalization—Members discount warnings and do not reconsider their assumptions.
3. Belief in inherent morality—Members believe in the rightness of their cause and therefore ignore the ethical or moral consequences of their decisions.
4. Stereotyped views of out-groups—Negative views of "enemy" make effective responses to conflict seem unnecessary.
5. Direct pressure on dissenters—Members are under pressure not to express arguments against any of the group's views.
6. Self-censorship—Doubts and deviations from the perceived group consensus are not expressed.
7. Illusion of unanimity—The majority view and judgments are assumed to be unanimous.
8. Self-appointed "mindguards"—Members protect the group and the leader from information that is problematic or contradictory to the group's cohesiveness, view, and/or decisions. (What is Groupthink?, n.d.)

For the past decade, educators have been under intense pressure to join the groupthink ideas of No Child Left Behind (NCLB) and the whole standards/accountability movement. Let's look at this list of traits and see how they fit today's education reform movement.

1. Illusion of invulnerability—Creates excessive optimism that encourages taking extreme risks.

Remember how NCLB was launched? We had the Texas Miracle, the high school dropout rates of zero reported by George W. Bush, which evaporated when it was revealed that the glowing stats were simply the result of administrative maneuvers and falsified data (Winerip, 2003). And the whole NCLB project projected that schools must reach 100% proficiency by 2014.

Promoters of charter schools have been claiming for years to have figured out how to overcome the effects of poverty. Only now that we are several years into the experiment do we hear that we must calibrate our expectations as they fail to deliver. We hear similar confident claims for the new Common Core State Standards (CCSS) tests, which will somehow magically wipe away the damaging effects of the previous tests. Not to mention the wonders of the Khan Academy and other computer-based delivery systems, which will allow us to simultaneously increase class sizes and personalize learning.

2. Collective rationalization—Members discount warnings and do not reconsider their assumptions.

We have had report after report documenting the failures of NCLB and high-stakes tests. Not one experiment in pay for test scores has worked (Hout & Elliott, 2011). Every time it fails, a reason is found that allows the idea to survive.

3. Belief in inherent morality—Members believe in the rightness of their cause and therefore ignore the ethical or moral consequences of their decisions.

What is the mantra of the phony reform movement—now repeated by Mitt Romney? "Education is the civil rights issue of our time." As if discrimination, housing, poverty, and voting rights no longer trouble us! Our schools are now more racially and economically segregated than any time since the 1960s, and this is given not a thought by these crusaders (Goldstein, 2011). Neighborhood schools are closed, entire staffs are fired, and dedicated teachers are subjected to humiliation by the press, all justified by this moral crusade for the children.

4. Stereotyped views of out-groups—Negative views of "enemy" make effective responses to conflict seem unnecessary.

Public education is defined as the status quo, and anyone who defends it is defending a failed, moribund system. There have been some interesting windows into the thinking of education reformers. A recent report on what members call the "Fight Club" reveals national coordination among various Education Reform Advocacy Organizations (ERAOs), such as Stand for Children, the Education Trust, and Students First (McGuinn, n.d.). They see as their enemy "the collection of teachers unions and other school employee associations derisively called the 'blob'," explains McGuinn. The unions are the chief villains in this morality play, acting to defend bad teachers, who morph into child molesters who cannot be fired or simply lazy individuals responsible for the economic decline of America.

5. Direct pressure on dissenters—Members are under pressure not to express arguments against any of the group's views.

This is most damaging within public schools when teacher evaluations are increasingly based on whether the teacher succeeds at embracing data-driven practices mandated from above. Administrators' careers advance, or not, depending on their willingness to distort and juggle data to support disastrous assessment-driven classroom practices.

In the public arena, Diane Ravitch is an archetypal heretic who was attacked when she left the conservative reform fold. Columnist Jonathan Alter (2010) called her the "Whittaker Chambers of school reform." Secretary of Education Arne Duncan said she was "in denial and she is insulting all of the hardworking teachers, principals and students all across the country who are proving her wrong every day" (Alter, 2011).

The current effort by NCTQ to silence criticism of standardized testing from schools of education is a frightening expansion of this campaign (Cody, 2012). When NCTQ's ratings of these schools are released, it will not be surprising to see schools of education that actively question the obsession over test-score data receive low scores, and preparation programs affiliated with alternative certification, such as Teach For America (TFA), receive high scores because of their devotion to data-driven instruction. And we will hear Secretary Duncan launch a program that removes funding from programs that produce graduates with lower test scores, as he has already indicated is planned. This is ideological and fiscal coercion, tying funding to data to punish those who deviate from the correct thinking.

6. Self-censorship—Doubts and deviations from the perceived group consensus are not expressed.

Within schools, there is pressure to join in the obsession over data, and this has intensified with recent reforms that require test scores to be used as a significant part of teacher and principal evaluations. Teachers who may have been willing to voice dissent in public in the past are now in fear of poor evaluations and possible termination. If one expresses a lack of faith in the latest curriculum or testing package, one might be accused of poor implementation or, worst of all, of the cardinal sin: not believing all students can learn.

A TFA corps member named James offered this advice in response to a post by TFA critic Gary Rubinstein (2012): "Corps members who choose to question TFA-doxy . . . should be prepared for an escalating series—in length—of 'mindset chats.'" I have not been privy to such a conversation, but clearly there is some heavy pressure at work to keep the corps members thinking a certain way.

7. Illusion of unanimity—The majority view and judgments are assumed to be unanimous.

Take a look at the 2010 multimedia extravaganza that accompanied the release of *Waiting for "Superman."* For Education Nation, the news division of NBC prepared a weeklong parade of education reform superstars such as Michelle Rhee, Bill Gates, Mike Bloomberg, and Cory Booker (Bernstein, 2010). There was a driving narrative that was almost unquestioned, with the exception of a hastily arranged Teacher Town Hall. This was the projection of a consensus where none exists. Anyone who disagreed with the main storyline was marginalized. Similarly propagandistic programming was aired on *Oprah* that week (Cody, 2010).

8. Self-appointed "mindguards"—Members protect the group and the leader from information that is problematic or contradictory to the group's cohesiveness, view, and/or decisions.

When President Obama accidentally spoke the truth a little more than a year ago and described an overemphasis on test scores, "using them to punish schools and students," the mindguards at the Department of Education leapt into the breach (Cody, 2011, March 29). They insisted that "[t]he President and Secretary Duncan are on the same page," though clearly the president's remarks were far different from his minion's policies. And when President Obama assembled a roundtable of advisers on education, not one educator was present (Strauss, 2011).

But in a bigger way, all of the organizations now being funded by the Gates, Broad, and kindred foundations are functioning as mindguards for the American public. We have Astroturf groups ready (Cody, 2011, July 16) to bring teachers to testify against their own due process protections, and groups such as Students First are willing to pour millions of dollars into lobbying policymakers to ensure they get the message about where their votes should go (Sawchuck, 2012). The Gates Foundation now funds the Media Bullpen, which has as its slogan "Bringing accountability to the media."

The site explains:

> The Bullpen reporters—the umpires—react and respond in real time to the press as it rolls out its coverage in print, online or broadcast, at the local, state or national levels. They score the coverage using the metaphors of that favorite American pastime, baseball. Articles are given strike outs, pop flies, singles, doubles, triples, and home runs, reflecting a particular story's objectivity, proper context, its exploration of data, and search for accuracy. (About the Media Bullpen, 2012)

While they claim to be diverse in their ideology, a review of the Bullpen's ratings reveals a strong bias in favor of standard education reform ideology. But like NCTQ, and Bill Gates himself, they attempt to position themselves above the fray, as umpires who set the rules of the game and determine who is in and out of bounds. They pretend to be beyond any particular ideology. "We are technocrats," Bill Gates recently said (Nocera, 2012). But technocracy and the set of solutions Gates and his experts have arrived at are deeply ideological, rooted in the mindset of the market, using test scores as the driving force in their quest to transform teaching. With the vast wealth of some of the world's most well-endowed foundations, they are purchasing the space where dialogue regarding education occurs.

The trouble with groupthink, as Janis points out, is that it can be disastrously wrong. Once we get swept up into this momentum, and more and more of our values and livelihoods hinge on this set of beliefs, it becomes harder and harder to break away. And with this particular set of ideas, we are, as a nation, building a huge technological infrastructure of curriculum, instructional tools, assessments, and data systems based on this diehard belief that test performance will drive learning to new heights. Those of us who have voiced skepticism are called Luddites or worse.

In cases like this, eventually the systems collapse, because reality will not support the endless optimism of the believers. The bubble always bursts, sooner or later. The NCLB testing bubble should have burst several years ago, and probably would have done so had not the billionaire technocrats intervened with the Common Core testing bailout. Now it looks like we are in for a few more years of glorious predictions of the wonderful equitable outcomes the latest and greatest testing technology will deliver, until it doesn't. But in the meantime, our public schools continue to be undermined, and resources continue to be diverted away from classrooms and into the testing/data infrastructure.

The sooner this groupthink bubble bursts, the better off we will be. In our classrooms, we must do our best to give students meaningful opportunities to learn, in spite of the intense pressure to raise test scores. In the public arena, we can help burst the bubble by focusing on the big-picture data that show that in spite of a decade of obsessing over data, there is no evidence that better learning results (Hout & Elliott, 2011). We can help burst the bubble by calling out the self-appointed umpires such as NCTQ, the Media Bullpen, and dozens of other test-obsessed advocacy groups that are attempting to overwhelm critical discussion of these issues. And we can support efforts to give voice to other points of view, through organizations that allow parents, teachers, and students to raise their voices, without the filtering effect of foundation funding.

References

About the Media Bullpen. (2012). The Media Bullpen. Retrieved from http://www.edweek.org/ew/articles/2012/05/16/31adv-foundations.h31.html

Alter, J. (2010, November 28). A case of senioritis. The Daily Beast. Retrieved from http://www.thedailybeast.com/newsweek/2010/11/28/alter-education-is-top-priority-for-gates.html

Alter, J. (2011, June 3). Don't believe critics: Education reform works. Bloomberg. http://www.bloomberg.com/news/2011-06-03/don-t-believe-critics-education-reform-works-jonathan-alter.html

Bernstein, K. (2010, September 18). The problem with NBC's Education Nation—Where are the voices of parents and teachers? Daily Kos. Retrieved from http://www.dailykos.com/story/2010/09/18/902996/-The-problem-with-NBC-s-Education-Nation-160-where-are-the-voices-of-parents-and-teachers

Cody, A. (2010, September 21). Oprahpaganda? Living in Dialogue [Web log post]. Retrieved from http://blogs.edweek.org/teachers/living-in-dialogue/2010/09/oprahpaganda.html

Cody, A. (2011, March 29). Obama blasts his own education policy. Living in Dialogue [Web log post]. Retrieved from http://blogs.edweek.org/teachers/living-in-dialogue/2011/03/obamas_radical_critique_of_tes.html

Cody, A. (2011, April 28). When will the testing bubble burst? Living in Dialogue [Web log post]. *Education Week*. Retrieved from http://blogs.edweek.org/teachers/living-in-dialogue/2011/04/when_will_the_testing_bubble_b.html

Cody, A. (2011, July 16). Teach plus: Astroturf in Indiana. Living in Dialogue [Web log post]. *Education Week*. Retrieved from http://blogs.edweek.org/teachers/living-in-dialogue/2011/07/teach_plus_astroturf_in_indian.html

Cody, A. (2012, May 25). Payola policy: NCTQ prepares its hit on schools of education. Living in Dialogue [Web log post]. Retrieved from http://blogs.edweek.org/teachers/living-in-dialogue/2012/05/payola_policy_nctq_prepares_it.html

Goldstein, D. (2011, January 17). On MLK Day, some thoughts on segregated schools, Arne Duncan, and President Obama. Dana Goldstein [Web log post]. Retrieved from http://www.danagoldstein.net/dana_goldstein/2011/01/on-mlk-day-some-thoughts-on-segregated-schools-arne-duncan-and-president-obama.html

Hout, M., & Elliott, S. W. (Eds.). (2011). Incentives and test-based accountability in education. Washington DC: The National Academies Press. Retrieved from http://www.nap.edu/catalog.php?record_id=12521

McGuinn, P. (n.d.). Fight club. *Education Next*. Retrieved from http://educationnext.org/fight-club/

Nocera, J. (2012, May 21). Gates puts the focus on teaching. *New York Times*. Retrieved from http://www.nytimes.com/2012/05/22/opinion/nocera-gates-puts-the-focus-on-teaching.html

Resmovitz, J. (2012, May 22). Education schools' training on standardized tests found lacking in new report. *Huffington Post*. Retrieved from http://www.huffingtonpost.com/2012/05/22/education-schools-standardized-tests_n_1536921.html

Rubenstein, G. (2012, May 17). Rusama Bin Stein? Gary Rubinstein's Blog [Web log post]. Retrieved from http://garyrubinstein.teachforus.org/2012/05/17/rusama-bin-stein/

Sawchuck, S. (2012, May 14). Foundation cash boosts education advocacy groups. *Education Week, 31*(31), 19. Retrieved from http://www.edweek.org/ew/articles/2012/05/16/31adv-foundations.h31.html

Strauss, V. (2011, July 18). President Obama's unusual education roundtable. The Answer Sheet [Web log post]. *The Washington Post*. Retrieved from http://www.washingtonpost.com/blogs/answer-sheet/post/president-obamas-unusual-education-roundtable/2011/07/18/gIQAf3UJMI_blog.html

What is groupthink? (n.d.). Psychologists for Social Responsibility. Retrieved from http://www.psysr.org/about/pubs_resources/groupthink%20overview.htm

Winerip, M. (2003, Fall). Houston's "zero dropout." *Rethinking Schools*. Retrieved from http://www.rethinkingschools.org/special_reports/bushplan/drop181.shtml

|| FOUR

Mean Scores in a Mean World

Lawrence Baines and Rhonda Goolsby

It was not so long ago that a visit from the state department of education was good news. In years past, the function of a state department of education was primarily "advisory, statistical, and exhortatory" (Johnston, 1999). Indeed, a hallmark of most successful schools is the help distributed through a powerful support system, such as a board or a Ministry of Education.

During his tenure as superintendent of Massachusetts schools, for example, Horace Mann often showed up unannounced in schools to observe teaching-in-action, to chat with children, and to ask what he could do to help. Mann preached the gospel of public schools as the antidote to problems of inequality, injustice, and the loss of civility (Baines, 2006). About using common schools to educate Americans of all classes and races, Mann (1867) said,

> If we are derelict in our duty in this matter, our children in their turn will suffer. If we permit the vulture's eggs to be hatched, it will then be too late to take care of the lambs. (p. 41)

The State as Vulture

Today, personnel from state departments of education are about as welcome in public schools as vultures. A wake of vultures seldom attacks healthy animals but prey upon the wounded or sick. So, when student achievement levels wane, the state sees its role not as helper, but as disciplinarian—to punish a school for allowing its students to post achievement scores below the mean. If a school is contacted by the state, the news inevitably is bad—at best, a public humiliation and, at worst, a tumult of teacher and administrator firings in a *takeover*. Firing people, while enjoyable for select politicians, is a tactic that helps neither student nor teacher.

Little wonder that teacher morale is at a 20-year low, and that one in three teachers admits that they are "very or fairly likely" to leave the classroom (Markow, Pieters, & Harris Interactive, 2012, p. 5). In many states, teachers are told not only what to teach but also how to teach by policymakers who have never set foot in their classrooms. Then, when test scores are reported as lower than expected, it is the teachers who get blamed—not the policies or policymakers.

Recently, we visited an elementary classroom in a low-income neighborhood that enrolled 30 students, a number well above the "suggested maximum enrollment" for young children of that age. The teacher was an energetic, endlessly patient, loving, smart, first-year teacher who expertly marched through lessons in reading and science. When writing a response to a prompt, one student broke her pencil and commenced crying. Over the next 60 minutes, she would cry again five more times—she did not understand a word, her eraser did not work properly, she needed to go to the restroom, someone nearby finished the assignment before she did, she discovered an ink stain on her hand.

Another student, who looked at least a year older than his classmates, kept trying to pick a fight with nearby peers, while a fidgeting autistic boy sat talking to himself and moving his head from side to side. Standing at the front of the classroom, and continually interrupting the teacher, was a new student from Guatemala who knew only a few words of English and tried to communicate by flailing his arms. A second new student sat silently at a desk in the back of class with his head on his desk. This student had just been placed with a new foster family after his father had been convicted on drug charges and sent to jail in a different city. These were only five of the young children in her class; there were 25 others.

Rather than help this first-year teacher in grappling with an incredibly challenging, large group of diverse children, the state has done absolutely nothing—provided no help in foreign language or special education, provided no relief from an overcrowded classroom, provided no mentor. Instead, the state has informed the teacher that her continued employment is dependent upon the test scores of these 30 students.

As countless education dollars (billions?) flow toward expensive, intricate evaluation systems, teacher pay is being slashed, class sizes are expanding, teacher tenure is sliding into oblivion, and real professional development is nonexistent. All funds in education today, without exception, must be somehow linked to test score ascension. Teachers who spend time caring for the emotional and social lives of children find themselves in the strange position of subverting the directives of the state.

Lean and Mean across the Nation

Among the lowest scorers in the recently made public teacher ratings for New York City were a group of accomplished, seasoned teachers from Public School 146 in Brooklyn (Winerip, 2012). Because 97% of their students had scored proficient in math the previous year, and only 89% scored proficient in math in the current year, these teachers were rated among the worst in the city. Yet, if only three children in their classes had scored one point higher on the exam, the teachers would have been rated average to better-than-average. Among this group of low-performing teachers was a Fulbright scholar, a former professor at Columbia, and a theater owner recognized by the Guggenheim Museum for developing an effective drama program for children.

In Florida, a state that rates each school and district by mean test scores, seesaw ratings have become so routine that the ratings have lost all credibility. A high school that receives an A one year might receive a C or D the next year, though the teaching and administrative staff may have remained completely intact from year to year (Florida Department of Education, 2012). One of the problems with rankings is that schools often go into testing blindly, not knowing the criteria by which they are going to be ranked. Although teachers are responsible for the delivery of instruction, testing and assessment are in the hands of policymakers.

By and large, the richest schools receive the highest scores and the poorest schools receive the lowest scores. As Tschinkel (2003) has noted in a series of articles decrying the inequity of the mathematical formula used to rate Florida's schools, "many schools serving a less affluent population have been graded lower than they deserve and many more affluent schools higher." About the ratings, Florida Teacher's Union president Randy Ford commented, "It's not that standardized test results don't tell us anything. They're very accurate measures of the size of the houses near a given school and the income levels of the people who live in those houses" (Postal, 2012). According to Rothwell (2012), "the average low-income student attends a school that scores at the 42nd percentile on state exams, while the average middle/high-income student attends a school that scores at the 61st percentile on state exams" (p. 1).

In Ohio, 80% of the highest poverty schools received a grade of either D or F in 2010–2011 while less than 1% of the richest schools received a D or an F (DiCarlo, 2012). None (as in zero) of Ohio's high-poverty schools were considered A-quality, while only 4% were considered B-quality. Meanwhile, 95% of the richest schools were rated either A- or B-quality.

As teachers know, some efforts and expenditures do not directly translate into higher test scores. Having a nurse onsite at a school, for example, used to be considered essential to ensure that children's health was adequately monitored, but the presence of a nurse, by itself, does not increase test scores. Thus, in most states, having a nurse on staff has become an extravagance that schools can no longer afford. In the state where I live, on average, there is only one nurse for every 3,110 students (Toppo, 2009).

Similarly, the quality of the school library and the work of the school librarian were once deemed integral to the basic functioning of a school. In the current era of reductionism, credentialed librarians have been fired and school library collections have withered. If a school actually has a librarian, he or she may have no training in research or library science. In California, for example, 76% of school libraries have no credentialed librarian, not even a part-time one. The ratio of students to librarians is 5,124 to 1 (California Department of Education, 2008) and the average copyright date of a nonfiction book in a California school library is 1972.

This ruthless, corporate-style focus on the bottom line has become a staple of a U.S. Department of Education that enthusiastically supports "state efforts to improve the quality of their assessment systems" (U.S. Department of Education, 2010, p. 11). With regard to budget cuts, the federal government admonishes that "it is important to do more with fewer resources" and warns that, in the future, funding will go only to schools "that are designed to significantly increase efficiency in the use of resources to improve student outcomes" (p. 41). In other words, don't expect the crises in school nurses or credentialed librarians to end anytime soon. All those ramshackle portable buildings that blight the grounds of half of the public schools in America are going to have to endure for another couple decades as well (U.S. Department of Education, 2011).

Recently, state legislators have cast their eyes on academics to streamline costs. Florida, South Carolina, Georgia, and Mississippi have taken the unusual step of forcing 15-year-olds to select a major for high school. According to the former governor of Florida, Jeb Bush, this edict allows students to "major in academic subjects such as foreign languages or history, or specific job areas such as auto mechanics. Regardless of a student's path, majors . . . help them understand the relevance between course work and their future" (Bush, 2006).

Bush's push to slot children into life paths at a young age has been in vogue outside the United States for years. In many European countries, it is common for students to be assigned to either the academic track or the trade track by age 15, based on their performance on a standardized test. For thousands of years, the Chinese have relied on a single, standardized test to help sort children into jobs suitable to their aptitudes and their station in life. Throughout much of the world, the richest students go to the best universities upon graduation from high school, while the poorest students enter the work force right away. Perhaps it is disingenuous for the United States to assert that it operates any differently.

The expense, uncertainty, and complexity of genuine human development make it an unpopular issue for many policymakers. In fact, for reformers of Bush's ilk, who perceive the function of a school to be preparation for future work (or a military career), human development is beside the point. Schools are just another business, susceptible to market supply and demand fluctuations, as encumbered by profits and debts as any business.

The Mean Kid

Consider the following three sets of scores.

Student A	Student B	Student C
100	70	100
100	80	80
100	70	70
0	80	50

All three sets of scores have a mean of 75, yet Student A looks as if he or she is capable of scoring the highest of the three whenever he or she wants to. But, what happened with that last score? Student B is fairly consistent, but does he or she really try? Student C started out strong, but recent difficulties are a cause for concern. Three different students with three distinctive stories, three diverse performances, and three different responses to learning (or, perhaps, testing). Yet all three students receive the same score with the expectation that all three should learn the same material.

Despite its crudity as a measure, the mean can be a malleable construct in the right hands. When we were teachers in Texas, for example, many schools used to routinely plan field trips for students in special education on testing day because the lower scores of special education students would crater the school mean. So, honest superintendents who did not plan a special field trip for their special education students and represented their scores accurately received lower mean scores. Their honesty, while laudable, placed at risk their jobs, the jobs of teachers and staff, and the school's reputation.

Similarly, recent research has revealed how KIPP (Knowledge Is Power Program) schools preselect out large numbers of students by requiring parents applying to the school to sign a pledge of involvement with their children and the school (Miron, Urschel, & Saxton, 2010). Furthermore, KIPP schools have the ability to kick out nonperforming or recalcitrant students, whereas public schools have no such option. As a result, in direct comparisons of schools with similar student populations, KIPP scores are almost always going to be higher because they have cut out their lowest-achieving students. Obviously, these students have to go to school somewhere. Inevitably, they show up at the nearest public school, thereby raising the mean scores at KIPP schools and lowering the mean scores at the public school.

Even when individual student performance is pried away from the group mean, what does a mean score reveal? If you were scouting the Boston Celtics during the 1964–1965 season in professional basketball, you would learn that the top nine players on the Celtics with the highest mean performance per minute of game time were as follows (Basketball Reference, 2012):

1. Ron Bonham, 26.7 points
2. Sam Jones, 25.8 points
3. John Havlicek, 22.8 points
4. Tom Heinsohn, 19.2 points
5. Willie Naulls, 18.4 points
6. Larry Siegfried, 16.4 points
7. Mel Counts, 16.2 points
8. Tom Sanders, 13.8 points
9. John Thompson, 11.8 points

The only problem with this "achievement list" is that it excludes Bill Russell, who only averaged 11.4 points for every 36 minutes of game time during the 1964–1965 season. Was Bill Russell a lesser player because his mean score was less than that of his teammates? For those who do not know or care about professional basketball, Russell was a five-time winner of the NBA Most Valuable Player Award, a 12-time All-Star, former captain of the U.S. Olympic Basketball Team, and considered by most sportswriters to be one of the greatest players in the history of professional basketball. He could score, but his forte was defense, rebounding, and a relentless work ethic.

Bill Russell's mean score did not provide a true indication of his contribution or talent. Similarly, students are being assessed on skills that are easily measurable—the standardized test equivalent of average points per minute. Students possess myriad talents and skills not addressed by narrow measures of content-area knowledge. The best teachers try to identify talent and give students time to work in areas of talent, as well as in areas that might need improvement. In valorizing test scores, a student's talent, individuality, and interests tend to be viewed as obstructive and oppositional to the goals of the state.

A Mean World

A common refrain of educational reformers in recent years has been the *unacceptable* scores of American students on the recent PISA (Program of International Student Assessment, 2010) reading test. About the 2009 PISA results, Secretary of Education Duncan (2010) explained, "Today's PISA results show that America needs to urgently accelerate student learning to remain competitive in the global economy of the 21st century. More parents, teachers, and leaders need to recognize the reality that other high-achieving nations are both out-educating us and out-competing us."

On the test, the average American 15-year-old (PISA only tests 15-year-olds) scored 500, 39 points behind top-scoring South Korea. Table 4.1 indicates the score of the average American in comparison with 15-year-olds in other countries.

Country	Score
1. South Korea	539
2. Finland	536
3. Canada	524
4. New Zealand	521
5. Japan	520
12. United States	500

Table 4.1. Top Five Scores on PISA Reading Assessment by Country

Using a mean score obscures a rather startling fact: many Americans are among the highest-achieving students on the planet. Students who have wealthy parents, for example, scored an average of 551 on the exam, which would make them the highest scoring group in the world. Just below rich Americans are Asian Americans and white Americans, who scored an average of 541 and 525, respectively, good enough for second and fourth best in the world (see Table 4.2).

Country	Score
1. Rich Americans (5,000,000 students of all races)	551
2. Asian Americans (all schools, 4,000,000 students)	541
3. South Korea (7,500,000 students)	539
4. Finland (850,000 students)	536
5. White Americans (all schools, 28,000,000 students)	525
6. Canada (5,500,000 students)	524

Table 4.2. Top Scores on PISA Reading Assessment by Country, with Wealth and Ethnicity Added for the United States

One can infer from these data that the quality of education received by 70% of Americans (the category being rich or white or Asian) appears satisfactory—good enough to place them among the highest-achieving students in the world. It is also useful to note that South Korea and Finland are relatively small, largely mono-cultural societies. Finland has fewer than a million students; South Korea has 7.5 million. In comparison, there are more than 50 million students in American public schools.

A related factor not considered in PISA or other international tests is the number of immigrant children included in testing. In South Korea the rate of immigration is 0. The immigration rate in Finland is .05, or about one person in 2,000. The few immigrants who come to Finland usually hail from Sweden, and both Swedish and Finnish are official languages of the country.

In contrast, the rate of immigration in the United States is 400 times the rate of immigration in Finland. In the United States, up to 21% of school-age children (ages 5–17) speak a language other than English at home and one in three schools in the United States have "official" immigrant populations of more than 25% (U.S. Department of Education, 2012). Unlike in some countries, in the United States, children of immigrants are free to enroll in public schools, even if they are unable to speak a word of English. American assessment systems require all students, irrespective of first language or the number of years spent in U.S. schools, to take the standardized test.

Yet another factor neglected in the calculation of mean scores is the 13% of American students classified as having special needs, meaning that they have been identified as having intellectual, emotional, or physical limitations. Special education is an American invention. Children with special needs in most other countries stay at home or attend a school designated for the disabled. In America, students in special education attend school with everyone else and are expected to take the same tests as everyone else.

Considering the additional challenges wrought by influxes of non-English-speaking students and the complications inherent in serving huge numbers of special education students on Individualized Education Plans (IEPs), Americans' world-leading scores are all the more impressive.

The Bad News

The bad news is actually not news at all but has become an ugly fact about education in the United States for a hundred years: Minority students who live in impoverished neighborhoods do poorly on standardized exams. On average, students who attend schools in America's poorest neighborhoods scored 3 points below children from Chile on the PISA Reading Assessment (see Table 4.3). American children who attend poor schools score an average of 446, 105 points behind American children who attend wealthy schools (and who lead the world at 551).

Students living in high-poverty neighborhoods in America who are high-achieving usually receive transportation to wealthier schools or have been the lucky recipients of dramatically increased funding. Indeed, Perry and McConney (2010) found the effects of socioeconomic status (SES) to be powerful and resilient across income groups. That is, "All students—regardless of their personal/family SES—benefit strongly and relatively equally from schooling contexts in which the SES of the school group is high" (Perry & McConney, 2010, pp.

1157–1158). Conversely, students performed markedly less well, irrespective of background, in low SES schools.

Country	Score
Austria	470
Turkey	464
Chile	449
United States (children from the poorest schools)	446
Mexico	425

Table 4.3. Bottom Five Scores on PISA Reading Assessment by Country, with Students in Poverty Added for the United States

Everyone's favorite example of the successful public school in a poor area seems to be Harlem Children's Zone, but the school is neither poor nor open to all. Harlem Children's Zone has an $84 million budget and assets of more than $200 million. The school receives $12,443 in public money and $3,482 in private money per pupil per year, but these costs do not include field trips, "a 4 p.m.-to-6 p.m. after-school program, rewards for student performance, a chef who prepares healthful meals, central administration and most building costs, and some of the expense of the students' free health and dental care" (Otterman, 2010).

While American schools are educating the middle class and the wealthy quite effectively, they have been woefully ineffective at educating poor African American and Hispanic children. In response to the unique needs of these minority students in poor schools, the federal government has pursued a policy of non-negotiable standardized testing accompanied by a funding model based on the percentage of students meeting minimal competencies. This regimen of high-stakes, low-challenge testing, initiated with Goals 2000 and continued with No Child Left Behind, has become a defining feature of Race to the Top, the Obama administration's signature education program.

Test Scores, Wealth, and Power

Standardized tests assess low-level knowledge in a specific subject area and ignore everything else. Once the standardized test becomes both curriculum and goal, its omnipresence subverts learning and makes the development of talent beside the point. But, even if the state decides that standardized test scores are what matters most, data indicate that 70% of American children are among the highest-scoring students in the world, despite the public schools' open doors to Limited English Proficient speakers and students with special needs. Everyone knows where the lowest-scoring children live and where they go to school. In America, the bottom 30% come from the poorest 30% of the population and are those most likely to drop out, be unemployed, and go to jail (Christensen, 2011). Yet, educational policy seems purposefully designed to reward the rich and to punish the poor.

Implicit in the rationale for inundating even our youngest children with tests is the contention that higher test scores will somehow make the United States more competitive globally. However, no evidence supports such a belief. The wealthiest countries in the world are listed in Table 4.4.

Purchasing Power Parity Ranking	Country	Purchasing Power (per person) 2010	Average PISA Score in Reading
1	Qatar	90,149	372
2	Luxembourg	79,411	472
3	Norway	52,964	503
4	United States	47,702	500
5	Switzerland	43,903	501
6	Netherlands	40,601	508
7	Australia	39,841	515
8	Austria	39,561	470
9	Canada	39,037	524
10	Ireland	39,009	496

Table 4.4. Wealthiest Countries in the World (Global Finance Magazine, 2012)

Few patterns of positive correlation exist between countries whose students do well on standardized exams and countries that are economically prosperous. History holds many examples that demonstrate that *school learning* has little impact on economic prosperity. The Roman Republic/Roman Empire, for example, with its massive population of slaves, high illiteracy rates, and inexorable military machine, ruled much of the planet for many years without the benefit of a formal, public system of education. The leaders of Rome focused on the accrual of power, without regard to human suffering or cultural degradation.

Similarly, in the United States today, power and the pursuit of wealth drive school reform. The goal is not human development, but the processing of children at the cheapest possible cost.

References

Baines, L. A. (2006). Does Horace Mann still matter? *Educational Horizons, 84*(4), 268–273.

Basketball Reference. (2012). 1964–65 Boston Celtics Roster and Statistics. Retrieved from http://www.basketball-reference.com/teams/BOS/1965.html

Bush, J. (2006, November 13). Should high school students be required to declare majors? *Upfront Magazine.* Retrieved from http://teacher.scholastic.com/scholasticnews/indepth/upfront/debate/index.asp?article=d1113

California Department of Education. (2008). Statistics about California school libraries. Retrieved from http://www.cde.ca.gov/ci/cr/lb/schoollibrstats08.asp

Christensen, L. (2011). The classroom to prison pipeline. *Rethinking Schools 26*(2). Retrieved from http://www.rethinkingschools.org/archive/26_02/26_02_christensen.shtml

DiCarlo, M. (2012, March 28). Ohio's new school rating system: Different results, same flawed methods. *Shanker Blog* [Web log post]. Retrieved from http://shankerblog.org/?p=5511

Duncan, A. (2010). Education Secretary Arne Duncan issues statement on the results of the Program for International Student Assessment. Retrieved from http://www.ed.gov/news/press-releases/education-secretary-arne-duncan-issues-statement-results-program-international-s

Florida Department of Education. (2012). Florida school district rankings. Retrieved from https://app2.fldoe.org/Ranking/Districts/

Global Finance Magazine. (2012). The richest countries in the world. Retrieved from http://www.gfmag.com/tools/global-database/economic-data/10501-the-richest-countries-in-the-world.html#axzz1stmRYW1Y

Johnston, R. (1999, June 23). State agencies take hands-on role in reform. *Education Week 18*(41), 1. Retrieved from http://www.edweek.org/ew/articles/1999/06/23/41power.h18.html

Mann, H. (1867). *The life and works of Horace Mann, Volume II.* Boston: Horace B. Fuller.

Markow, D., Pieters, A., & Harris Interactive. (2012). *Survey of American teachers: Teachers, parents and the economy.* New York: MetLife. Retrieved from http://www.metlife.com/about/corporate-profile/citizenship/metlife-foundation/metlife-survey-of-the-american-teacher.html?WT.mc_id=vu1101

Miron, G., Urschel, J., & Saxton, N. (2010). *What makes KIPP work? A study of student characteristics, attrition, and school finance.* New York: National Center for the Study of Privatization in Education. Retrieved from http://www.ncspe.org/publications_files/OP195_3.pdf

Otterman, S. (2010, October 12). Lauded Harlem schools have their own problems. *New York Times.* Retrieved from http://www.nytimes.com/2010/10/13/education/13harlem.html

Perry, L., & McConney, A. (2010). Does the SES of the school matter? An examination of socioeconomic status and student achievement using PISA 2003. *Teachers College Record 112*(4), 1137–1162.

Postal, L. (2012, January 30). Educators criticize latest Florida school rankings. *Orlando Sentinel.* Retrieved from http://articles.orlandosentinel.com/2012-01-30/news/os-florida-school-rankings-20120130_1_half-on-fcat-scores-fcat-scores-and-half-school-rankings

Program of International Student Achievement. (2010). PISA 2009 results: What students know and can do: Student performance in reading, mathematics and science. Paris, FR: OECD. Retrieved from http://dx.doi.org/10.1787/9789264091450-en

Rothwell, J. (2012). Housing costs, zoning, and access to high-scoring schools. Washington, D. C.: Brookings Institute. Retrieved from http://www.brookings.edu/~/media/Files/rc/papers/2012/0419_school_inequality_rothwell/0419_school_inequality_rothwell.pdf

Toppo, G. (2009, August 10). School nurses in short supply. *USA Today.* Retrieved from http://www.usatoday.com/news/health/2009-08-10-school-nurses_N.htm

Tschinkel, W. (2003). New and improved A+ grades: Camouflaged bias. *School performance articles.* Retrieved from http://www.bio.fsu.edu/~tschink/school_performance/

U.S. Department of Education. (2010). *A blueprint for reform.* Washington, DC: Government Printing Office.

U.S. Department of Education. (2011). Percentage of public schools with permanent and portable (temporary) buildings and with environmental factors that interfere with instruction in classrooms, by selected school characteristics, type of factor, and extent of interference, 2005. *Digest of Education Statistics, 2010.* Washington, DC: Government Printing Office. Retrieved from http://nces.ed.gov/programs/digest/d10/tables/dt10_106.asp

U.S. Department of Education. (2012). Table A-6-2. Number and percentage of children ages 5–17 who spoke a language other than English at home and who spoke English with difficulty, by age and selected characteristics: 2009. *The Condition of Education, 2011*. Retrieved from http://nces.ed.gov/programs/coe/tables/table-lsm-2.asp

Winerip, M. (2012, March 4). Hard-working teachers, sabotaged when student test scores slip. *New York Times*. Retrieved from http://www.nytimes.com/2012/03/05/nyregion/in-brooklyn-hard-working-teachers-sabotaged-when-student-test-scores-slip.html

|| FIVE

De-Grading Literacy

How New York State Tests Knowledge, Culture, and Critical Thinking

Julie A. Gorlewski and David A. Gorlewski

Setting the Stage for De-Grading

In June 1999, New York State anticipated the political and pedagogical movement that has engulfed public schools through the federal legislation titled No Child Left Behind (USDE, 2003). The state's education department implemented learning standards meant to drive local district curricula. In addition, the state unveiled a plan to attach the standards to mandatory assessments for students in grades 4, 8, and 11, beginning in the area of English language arts (ELA). Consequences for students and educators were significant and comprehensive. In addition to gauging individual student performance, tests at all levels were designed to measure schools' progress toward meeting the learning standards and to rank schools according to student achievement. Scores and rankings were to be published and distributed by districts, the state education department, and media outlets; schools with consistently inadequate scores and unacceptable levels of improvement were threatened with the designation "School Under Regents Review (SURR)." So-called SURR schools would be required to show rapid, significant improvement on standardized assessments or face state takeover (NYSED, 1999). Tests were equally high-stakes for students. In June 1999, passing the commencement level ELA examination (intended for students in grade 11) became a graduation requirement for the high school graduating class of 2000.

Couched in the rhetoric of equity and accountability, these reforms are consistent with what Michael Apple (1996) describes as a movement toward a national curriculum:

> While the proponents of a national curriculum may see it as a means to provide social cohesion and to give all of us the capacity to improve our schools by measuring them against "objective" criteria, the effects will be the opposite. . . . Rather than leading to cultural and social cohesion, differences between "we" and the "others" will be produced even more strongly, and the attendant social antagonisms and cultural and economic destruction will worsen.

Furthermore, a national curriculum provides a controlled structure in which a national testing system can operate. By offering *objective accountability measures* in the form of state-approved and -administered assessments, schools can be perceived as products in an educational marketplace where *consumers* have both the freedom to choose and the information on which to base their decisions. Unfortunately, this scenario disregards issues of inequality since school quality is rated as though district resources were identical. Proponents proceed as though implementation of standards and assessments, coupled with public scrutiny, will somehow result in academic improvement. Instead, standardized assessments reveal existing social inequities, and then deem poor performing school systems deficient.

Bernstein (1971) explicates associations between curriculum, instruction, and assessment: "Curriculum defines what counts as valid knowledge, pedagogy defines what counts as a valid transmission of knowledge, and evaluation defines what counts as a valid realization of knowledge on the part of the taught" (p. 85). He notes the inextricable, recursive connections among curriculum, pedagogy, and evaluation. Bernstein's analysis illustrates the *constructed* nature of knowledge by illuminating the fact that curriculum, pedagogy, and evaluation define what counts as knowledge. This point of view is not consistent with the rhetoric around public school reform as described by Apple (2001). He notes that education is seen (simply) as the delivery of neutral knowledge and that the effectiveness of the delivery is best measured by standardized achievement tests. These conceptions contribute to the degradation of knowledge through the regimes of standardized testing.

Standardized Testing

> What gets measured gets taught. What gets reported gets taught twice as well. (Rothman, 2001)

Many scholars have argued that it is through evaluation that definitions of knowledge become realized. Smith (1986) synthesized 20 years of educational research on instruction and assessment to determine how standardized testing affects learning. He states: "Two assumptions underlie the 'quality control' of continual testing: (1) that children will learn more and (2) that teachers will teach better" (p. 129). He notes that each assumption is flawed. Although he wrote years prior to NCLB legislation, Smith highlights the political and value-based aspects of standardized tests: "It is through testing that politicians, administrators, and supposedly omniscient outsiders exert the control that they expect will achieve excellence in education" (p. 130).

Similarly, Popham (2001) states: "In an evidence-oriented enterprise, those who control the evidence-gathering mechanisms control the entire enterprise" (p. 130). Popham reviews the history of testing practices in education and addresses the ways in which using high-stakes tests to measure school accountability changes the essential questions on which educators focus. He describes this shift as follows: "The critical question of 'How do we teach Tracy the things she needs to know?' is forced aside by this far less important one, 'How do we improve Tracy's scores on the high-stakes test she will be taking?'"

This is relevant in that the basis of assessment—which is the driving element of definitions of knowledge—has "progressed" from student/teacher, to school, to district, to state, to federal control.

Kohn (2004) explicates the consequences of this progression by providing an analysis of the effects of accountability reforms that center on standardized tests. Kohn describes the political reasons that underlie the federal No Child Left Behind (USDE, 2003) legislation, and he ties those reasons to economic purposes, primarily the privatization of public education. Kohn offers a large-scale connection between standardized tests taken by individual students and the social, political, and economic forces that support them, stating:

> Consider these examples of what I'm calling collateral damage from high-stakes testing: a more traditional, back-to-basics curriculum; more homogeneity; a retreat from innovations like multi-age classrooms; more tracking and retention and harsher disciplines. What's striking about these ostensibly accidental byproducts of policies designed to ensure accountability is that they, themselves, are on the wish list of many of the same people who push for more testing—and, often, for vouchers. (p. 13)

Kohn's assertions are reminiscent of Apple's (1996) prediction that a standardized curriculum would have effects that are not cohesive, but divisive. It is evident that using tests as central measures of educational accountability is problematic on several fronts. Therefore, an examination of the connections between institutional instructional accountability and the political question of "who decides" the bases of evaluations that determine progress toward it might begin with an analysis of the assessments themselves.

In New York State, education department officials decided to begin their reform initiative in the curriculum area of English language arts (ELA). While there is no record of their reasoning, it is likely that the lack of specific "content" in this subject area was relevant. Therefore, reformers could focus on the "how" of assessment, reserving disciplines in which the "what" of assessment would prove problematic. ELA standards illustrate this emphasis by listing *four* language arts skills (reading, writing, listening, and speaking) and *four* purposes for those skills (language for information and understanding; for literary response and expression; for critical analysis and evaluation; and for social interaction). They are devoid of content.

Criteria by which these standards might be assessed are articulated by Tucker and Codding (2001):

> Standards should be usable by students, in the sense that a student should be able to look at the standards and know instantly what topics have to be mastered, what knowledge has to be gained, and what kind of work he or she has to produce to meet the standard. By the same token, standards should be usable by teachers, in the sense that teachers should be able to look at the standards and know what topics they have to teach, what the students need to know, and what kind of work their students have to do to meet the standards. (p. 479)

Clearly, these standards fall short of the criteria set by Tucker and Codding. They are general, conceptual, and offer no specificity. The language of the standards seems politically and academically unassailable. They appear to allow for an array of possibilities regarding definitions of knowledge.

Although English language arts offered policymakers a relatively *content-free* discipline in which to begin its testing program, the area of *language* encompasses significant matters of thought, culture, and identity. Hillocks (1995) explains how writing shapes thinking. In addition, he clarifies the connection between teaching writing and facilitating students' use of language for discovery and self-expression. Moffett (1988) discusses the relationship between writing and personal empowerment. He asserts that writers are empowered by literacy. Writing hones their voices and authorship strengthens the meanings in their messages.

If, as Hillocks (1995) asserts, writing "shapes thinking," then students' experiences with, and practices of, writing affect who they become and how they will represent themselves and their cultures through discourse. Moffett (1988) points out that "[i]t is impossible to understand the teaching of writing in America if one does not realize that, in one form or another, from first grade through graduate school, it serves mostly to test reading" (p. 86). This contributes to the fact that writing is perceived as "answering questions"; it is virtually never viewed in schools as a way of learning or thinking.

In this complex consideration of language arts, Hillocks (2002) defines the problem of testing by asking: Does a writing test assess what its proponents claim? Does it indicate how well a student may be able to write in any given situation? Are these tests meaningful measures of individuals and schools? In addition, he centers on inconsistencies between assessments and the standards they are meant to reflect.

Hillocks (2003) recommends "assessing the assessments" as a strategy for minimizing educational deficiencies that result from a preoccupation with standardized writing tests. He justifies this approach based on an exhaustive analysis of standardized state writing assessments:

> The chief finding of this study is that writing assessment drives instruction. It stipulates the kinds of writing that should be taught; it sets the standards for what counts as good writing; and it sets the conditions under which students must demonstrate their proficiency, and, as a result, sets out what students learn. (p. 67)

Since institutional and pedagogical influences of the curriculum are reinforced by the regulations and publications of the state, it is important to analyze examinations to determine definitions and applications of knowledge that are privileged within them.

Method

This chapter provides data and analyses of two sets of New York State Education Department examinations representing grade 11 and grade 4.

Grade 11: The commencement level grade 11 language arts examination, referred to as the New York State Regents Examination, was first administered in 1999. The analyses encompass examinations administered from June 1999 through August 2004. Seventeen standard examinations were administered during the period of analysis. Analytic criteria for the examination included the following:

- The extent to which the exam considers states of knowledge and/or ways of knowing;
- What counts as knowledge, that is, topics included for reading passages, genres incorporated for both reading and writing, and criteria provided for student test takers;
- The type of knowledge tested, that is, the extent to which the test questions assess critical thinking;
- Level of curricular discipline imposed by the examination; and
- Theory of writing espoused by the examination.

The final criterion is essential because the theory of writing can be linked to the values that state education department assessors place on students' thinking and learning. As Hillocks (2003) states, "The initial decision about what counts as writing has powerful implications about what will be tested" (p. 8).

Grade 4: The grade 4 English Language Arts Test for 1999 and 2000 represented the first two administrations of this new testing instrument. They both consisted of 28 multiple-choice/cloze questions related to reading comprehension skills, four short responses, and three extended responses. Both were timed tests scheduled to be completed within 165 minutes over a three-session period. Analytic criteria for the examination included the following:

- The extent to which the individual questions represent higher order thinking skills as reflected by Bloom's Taxonomy. In essence, this represents an analysis of the cognitive domain.

Results—Grade 11 Examination Analysis

In general, the state's assessment system has failed to address the elevated ambitions espoused in its own standards. Instead, writing is used in a strictly controlled structure to re-inscribe the authority of dominant texts. State regulation is prevalent, from the compulsory nature of the exam, to procedures of administration, to writing prompts and guidelines.

New York State's commencement examination is composed of four tasks to be completed in two sessions (3 hours each) on consecutive days. Every element of examination administration is prescribed: directions are scripted for teachers to read; exams must be given only during the time allotted; at completion, both students and teachers must sign oaths attesting to the integrity of exam administration. Exams may not be opened from sealed wrappers prior to the designated starting times, and answer keys and anchor papers may not be reviewed by teachers until students have finished writing. Before exams are even begun, much less scored, the authority and control of the state education department is conspicuous and palpable.

The scoring procedure is equally regulated. The state document *Information Booklet for Administering and Scoring the Regents Comprehensive Examination in English* is 16 pages long and includes two scoring grids that must be photocopied and completed by all participating scorers. The scoring keys (which are specific to each exam) are voluminous; none has fewer than 100 pages. Each of four student essays must be graded by two different scorers. Scoring directions require that districts make every effort to avoid having teachers rate their own students' work.

The discourse of control, with its consequential reduction of writing assessments to numerical values, gives the impression of operational efficiency and accuracy. For each exam administration, scorers (teachers) must be retrained with new rubrics and scores are determined through a new "scoring scale." The procedure reeks of accuracy, but the precision is a pretense, since the exams assess a minute fraction of what people experience as language arts; and, in fact, the state exam addresses little of what its own standards claim is important.

Results of the analysis fell into three main areas: First, considerations of race, ethnicity, and gender; second, the nature of knowledge; and third, what counts as student writing.

Erasing Gender, Race, and Ethnicity

Analysis of the mandatory New York State ELA Regents Examination revealed gender, racial, and ethnic related biases. In fact, even the most superficial investigation exposed significant gender bias. Each exam includes six passages of text and one "critical lens" (based on a quotation to which students must relate two works of literature from previous readings). All passages were coded by author's gender; where gender could not be determined, the passage was coded as "no gender indicated." Furthermore, in some passages, content included references to gender. For example, one passage dealt with *the nature of boyhood friendships* and was coded "male." Another passage dealt with *mother and daughter relationships* and was coded "female." In some cases, multiple coding was necessary. A relevant example involves a writing prompt about *What mothers do for their children.* This piece was coded "female" because the topic related to mothers; however, both associated passages were written by male authors. Coding results revealed a significant difference in the gender of authors represented on the exams. Considering strictly authorship, 18 texts were undeterminable, 20 texts were authored by females, and 62 were authored by males. Another salient point on this issue involves "who speaks" or "who is knowledgeable" on what kinds of issues. Male authors wrote about a wide range of topics, including science, history, politics, sports, relationships, the arts, and gender-related issues regarding both men and women (i.e., men wrote about mothers *and* boyhood friendships). On the other hand, women's topics were limited in both volume and content. Female authors included on the exam wrote about history, literacy, relationships, personal growth, and childhood/adolescence. Women addressed gender-related issues only when they were associated with women, that is, women wrote about the struggle for women's suffrage and mother-daughter relationships. Men wrote freely about a variety of issues. Male authors wrote about fathers, mothers, daughters, and sons. Moreover, males wrote as authorities about women, as demonstrated when two male-authored texts were used as source documents for the topic of "the influence of mothers" and when two male-authored texts about their respective elementary school teachers were used as source documents for the topic, "how teachers influence students." In both cases, male-authored texts reinscribed women in traditional roles. In contrast, women's texts were limited and did not speak about men.

The exam also minimizes diverse perspectives about gender, race, and ethnicity. Both the New York State Civil Liberties Union (NYSCLU) and the National Coalition Against Censorship (NCAC) have denounced the state education department for editing passages on the ELA regents examination (NCAC, 2002;

NYSCLU, 2002). Homosexuality is erased from Annie Dillard's passage about the surprises involved in developing characters (censored elements are bold):

From Annie Dillard's *Bird by Bird* (which appeared in the January 2001 exam):

> If you can get their speech mannerisms right, you will know what they're wearing and driving and maybe thinking, and how they were raised, and what they feel. You need to trust yourself to hear what they are saying over what you are saying. At least give each of them a shot at expression: sometimes what they are saying and how they are saying it will finally show you who they are and what is really happening. Whoa—they're not getting married after all! **She's gay!** And you had no idea! (NCAC)

The reference to a character being gay was removed without Dillard's permission. Literary references to oppression on the basis of race, ethnicity, and gender have been deleted from other passages as well:

From B.B. King's autobiography, *Blues All Around Me* (June 2000 exam):

> My great-grandmother, who'd also been a slave, talked about the old days. She'd [She would] talk about the beginnings of the blues. She said that, sure, singing helped the day go by. Singing about sadness unburdens your soul. But the blues hollerers shouted about more than being sad. They were also delivering messages in musical code. If the master was coming, you might sing a hidden warning to the other field hands. Maybe you'd want to get out of his way or hide. **That was important for the women because the master could have anything he wanted. If he liked a woman, he could take her sexually. And the woman had only two choices: Do what the master demands or kill herself. There was no in-between.** The blues could warn you what was coming. I could see the blues was about survival. (NCAC)

From Annie Dillard's memoir, *An American Childhood* (August 2001 exam):

> From the nearest library, I learned every sort of surprising thing—some of it, though not much of it—from the books themselves.
>
> The Homewood branch of Pittsburgh's Carnegie Library system was in a Negro section of town—Homewood. This branch was our nearest library; Mother drove me to it every two weeks for many years, until I could drive there myself. I only very rarely saw other white people there.
>
> Beside the farthest wall, and under leaded windows set ten feet from the floor, so that no human being could ever see anything from them—next to the wall, and at the farthest remove from the idle librarians at their curved wooden counter, and from the oak bench where my mother waited in her camel's-hair coat chatting with the librarians or reading—stood the last and darkest and most obscure of the tall nonfiction stacks: **NEGRO HISTORY and** NATURAL HISTORY. (NCAC)

The erasing of "potentially offensive" references to race, ethnicity, and gender, while at the same time privileging the "natural and normal" dominance of white male knowledge, is integral to the content of the examination.

What Counts as Knowledge

As listeners and readers, students will collect data, facts, and ideas; discover relationships, concepts, and generalizations; and use knowledge generated from oral, written, and electronically produced texts (NYSED, 1996).

The structure of the exam indicates that knowledge is provided by authorities and not constructed by learners. The exam is divided into four tasks, each of which involves writing. In the first task, students must listen to a passage, answer multiple-choice questions, and then respond to a prompt regarding the passage. The second task has students read a piece of nonfiction ("information" oriented, not memoirs or autobiographies) and interpret a related "graphic" (i.e., chart, map, or table), answer multiple-choice questions about each, and then respond to a writing prompt that connects both. The third task (day two of the exam) involves two "literary" passages that are meant to connect thematically to a "controlling idea." Again, students answer multiple-choice questions about each work and then respond to a writing prompt. Task four is termed the "critical lens." A critical lens is a quotation through which literature might be interpreted. The first exam, in June 1999, contained the following critical lens: "In literature, evil often triumphs but never conquers." Task four directions instruct students to "provide a valid interpretation" of the critical lens and then relate it to two works of literature "using specific references to appropriate literary elements."

A review of passage topics is noteworthy. A desire to appeal to youth interests is evident in the inclusion of subjects such as "teen sleep patterns" and "work-related teen injuries," but other patterns are apparent as well. Some topics that emerged repeatedly included civics, the environment, schooling, relationships, and science/technology. *Progressive* positivistic science is privileged in tasks like the one in April 2004, when students were instructed to "explain why some people think that eventually we will have a manned space mission to Mars." Technology is privileged similarly; in January 2001, students were directed to "discuss the merits and possible uses of free public access to computer networking." In January 2002, the directions read: "You have decided to write a letter . . . in which you persuade the committee to provide funding for the use of technology to save endangered animals."

Another recurring theme involves business and management. Several exams require students to engage with passages in which business strategies are esteemed. In the August 2003 exam, students were instructed to listen to a speech about Sam Walton and then "write a proposal . . . explaining how you would use business strategies to plan a successful fund raising campaign." The January 2003 exam requires students to listen to a speech by former Pittsburgh Steelers coach Chuck Noll and then "prepare a report on successful management techniques in the workplace." The notion that public institutions can be improved by emulating capitalistic private sector methods is prevalent (Apple, 2001; Kohn, 2004), although as yet unproven. The tendency of the state to reproduce market ideology through text on compulsory examinations reinforces the sense that standardization will favor the privatization and marketization of public education (Apple, 1996, 2001).

Furthermore, although decades of research indicate the connection between writing proficiency and choice (Atwell, 1987; Hillocks, 1986; Kitagawa & Kitagawa, 1987; National Writing Project & Nagin, 2003; Tharp & Gallimore, 1988), the state assessment provides minimal opportunity for students to employ choice as writers. Topics and passages are prescribed, and students' positions on topics are predetermined three times as often as not. For example, students are offered a choice of position in August 2000, when the prompt reads: "For a class debate, your teacher has asked you to write a position paper indicating whether you agree or disagree that the ratings system will help parents and guardians monitor their children's television viewing."

Political bias that privileges patriotism and minimizes student agency is revealed as well. On one test, students are directed to listen to a passage about the history of the national anthem and then write an essay arguing that *it should be maintained* as such. In the 17 exams reviewed, each of which requires four tasks, such a "choice" is provided eight times. A more typical example of prescriptive prompts is reflected in June 2001:

> The planning team in your school is interested in developing a school-to-work program in conjunction with businesses in the community. As a member of the Career Guidance class, you have been asked to write a letter to the planning team in which you describe the benefits of school-to-work programs and the conditions needed to make such programs successful.

Ironically, the rhetoric of task one in the August 2000 examination *implies* student choice without actually providing it: "You have decided to write a letter to the editor of your local newspaper arguing for the importance of an education in the arts."

It is clear that the New York State ELA Regents examination supports the perception of knowledge as materializing from external authorities and comprising sanctioned, inoffensive topics. Moreover, student choice is exceedingly restricted, reserved to two literary selections that must relate to the critical lens task, and less than one-third of the remaining writing prompts. Most aspects of this examination, which purports to measure proficiency in the area of English language arts, have little to do with the development or performance of authentic discourse.

What Counts as Student Writing

The initial decision about what counts as writing has powerful implications for what will be tested (Hillocks, 2003).

Although the state's learning standards promote higher order thinking and using language to develop understanding, its assessment implements a narrow stance regarding what counts as student writing. The "authoritative" texts sanctioned through state publications exhibit a variety of genres and styles; however, student writing is strictly limited. This format is not unusual. According to Hillocks (2003), most state writing assessments have similar limitations: "Students are asked to explain, interpret, or describe something based on background experiences or information provided in the prompt" (p. 67). New York State's ELA exam does not include students' writing about background experiences, except insofar as they relate to having read a piece of literature that links to the "critical lens" task.

Writing prompts control the topic, the information used to develop student responses, the content of the responses, and (quite often) the position students must take on the issue at hand. The directions and guidelines for the January 2004 critical lens, arguably the task in which students have the greatest opportunity to exercise choice, exhibits rigorous control:

> Your Task:
>
> Write a critical essay in which you discuss ***two*** works of literature you have read from the particular perspective of the statement that is provided for you in the **Critical Lens**. In your essay, provide a valid interpretation of the statement, agree ***or*** disagree with the statement as you have interpreted it, and support your opinion using specific references to appropriate literary elements from the two works. You may use scrap paper to plan your response. Write your essay in Part B, beginning on page 9 of the essay booklet.
>
> Critical Lens:
>
> "Things can happen in some cities* and the tale of them will be interesting; the same story laid in another city* would be ridiculous." —Frank Norris, *McTeague: A Story of San Francisco: Authoritative Text, Contexts, Criticism,* 1997
>
> * For the purpose of writing your critical essay, you may interpret the word "cities" to mean locations and the word "city" to mean location.
>
> Guidelines:
>
> Be sure to:
>
> - Provide a valid interpretation of the critical lens that clearly establishes the criteria for analysis

- Indicate whether you agree ***or*** disagree with the statement as you have interpreted it
- Choose ***two*** works you have read that you believe best support your opinion
- Use the criteria suggested by the critical lens to analyze the works you have chosen
- Avoid plot summary. Instead, use specific references to appropriate literary elements (for example: theme, characterization, setting, point of view) to develop your analysis
- Organize your ideas in a unified and coherent manner
- Specify the titles and authors of the literature you choose
- Follow the conventions of standard written English

The perception of student writing promoted by this exam is detrimental to teaching, learning, and thinking. With its emphasis on prescriptions, Moffett (1988) describes this as "formalism" (p. 87) that pervades student writing, resulting in what he calls "decomposition" (p. 87).

As Moffett (1988) noted, writing is primarily used to assess reading. Twenty-four years later, his assertion holds true. On the New York State ELA exam, students' writing is used to assess neither writing ability nor higher order thinking skills. Instead, it is used primarily as a measure of reading comprehension. Students are required to read, answer multiple-choice questions about texts, and then summarize key points of texts according to a particular writing prompt. Tasks are neither authentic nor meaningful; they offer none of the elements that foster excellence in writing (Goodman & Goodman, 1990; Hillocks, 1986; National Writing Project & Nagin, 2003). Unfortunately, these assessments are likely to influence curriculum and instruction in public schools throughout New York State.

Results—Grade 4 English Language Arts Test

This analysis attempted to *quantify* the cognitive domain by placing a numeric value on each rung of Bloom's Taxonomy and then determining a specific thinking skill level for each question on the grade 4 English language arts test in 1999 and 2000.The premise in the use of the taxonomy is that educational objectives can be arranged in a hierarchy from less to more complex. There are six thought process levels in the taxonomy: knowledge, understanding, application, analysis, synthesis, and evaluation. Each of these levels is manifested by the existence of indicative actions. For example, the level "knowledge" would be tested if a student could list, name, or describe something. Therefore, a question that asks a student to list or describe something would be a question that tests a student's knowledge. On the other end of the taxonomy is "evaluation," which is manifested by actions such as make a choice and justify, decide how and support, and select, and why. Questions such as these test the student's ability to evaluate.

A Bloom's Taxonomy Rating Scale was developed to rate each test question. To facilitate categorizing and ranking, each level of the taxonomy was assigned a numerical value from one to six, as follows: knowledge (1), understanding (2), application (3), analysis (4), synthesis (5), and evaluation (6). The numerical value was based on a modification of the Rigor/Relevance Framework (Daggett, 1999). The resulting framework encompasses two types of learning: knowledge within a discipline (Bloom's Taxonomy) and the ability to use that knowledge to solve real-life problems (Application Model).

After each question from the 1999 and 2000 tests was given a numerical rating, the numbers were totaled and then divided by the number of questions to establish a Bloom's Taxonomy Rating for each test.

Despite New York State Education Department reform documents purporting to represent the adoption of new standards, greater rigor, increased graduation requirements, and new accountability measures (in the form of School Report Cards), the questions on the 1999 test had a total Bloom's Taxonomy Rating of 1.86 while the 2000 test was rated at 1.71. In terms of Bloom's Taxonomy, the average thinking skills level for both tests was between *knowledge* and *application*—the two lowest rungs on the taxonomy.

The low taxonomy rating, coupled with multiple-choice and short response questions, are indicative of the following: First, the fourth-grade New York State ELA tests are *not designed* to test the upper rungs of Bloom's Taxonomy (such as analysis, synthesis, and evaluation). Second, the tests *are designed* to be evaluated quickly and efficiently. By adopting a testing format that included multiple-choice questions and questions requiring brief responses, and by mandating specified time frames for administration and completion, the state ensured that reform initiatives could be easily monitored and that data could be generated to compare students, schools, and school districts.

De-Grading Knowledge through Standardization

Apple (1996) predicted the effects of a national curriculum as detrimental for public education. The analyses of New York State tests at the grade 11 and grade 4 levels revealed a range of concerns relative to format, rigor, and concepts of knowledge. New York State's Regents Examination, as an early illustration of NCLB's intended outcome, upholds Apple's concerns. The exam, while professing objectivity and equity, erases any examples of oppression and bias within the culture. It explicitly privileges white male writers. Simultaneously, references to race, gender, and homosexuality are expunged from passages without the permission of the authors. Moreover, men can speak on an unlimited array of issues, while women are restricted to traditional feminine topics such as family, literacy, and women's rights.

In addition, "progressive" values of science and technology are privileged, as is patriotism. Other recurrent topics include schooling, the arts, the environment, civics, family, and teen-related issues. Business-oriented strategies are esteemed as models for students to emulate.

Finally, student writers are perceived as responsive, not expressive. The state examination provides the writing prompt, relevant information, "controlling ideas," and (in most cases) the position students must take on an issue. This perspective reinforces the point of view that students are recipients of state-issued knowledge, and that writing is merely the means to assess their compliance as consumers of information in a politically regulated system of education.

The grade 4 test analyses, on the other hand, showed deficiencies of a different kind. Primarily, students were required to respond to questions that were not, in any way, related to higher order, that is, critical thinking. Despite the state's contention that these tests represented fundamental reform and elementary level preparation for 21st-century challenges, a Bloom's Taxonomy rating for each test question and for each test indicated that the questions posed required student facility in only the lowest rungs (knowledge and application) of the taxonomy. If the state wanted to achieve rigor through these tests, it fell far short of that goal.

In addition, the grade 4 tests, consistent with the new reporting system, were designed for ease of scoring and public reporting. The test format generated data that enabled the state to rank schools and school districts and publish the results for public consumption. It should be noted that these efficiencies were achieved, in a sense, by *degrading* the cognitive level of the examination.

References

Apple, M. W. (1996). *Cultural politics and education*. New York: Teachers College.

Apple, M. W. (2001). *Educating the "right" way: Markets, standards, God and inequality*. New York: Routledge-Falmer.

Atwell, N. (1987). *In the middle: Writing, reading, and learning with adolescents*. Upper Montclair, NJ: Boynton/Cook.

Bernstein, B. (1971). On the classification and framing of educational knowledge. In M. F. D. Young (Ed.), *Knowledge and control: New directions for the sociology of education* (pp. 47–69). London: Collier-Macmillan.

Daggett, W. (1999). *Rigor/Relevance Framework*. New York: International Center for Leadership in Education. Retrieved February 12, 2000, http://www.leadered.com/rrr.html.

Goodman, Y. M., & Goodman, K. S. (1990). Vygotsky in a whole language perspective. In L. C. Moll (Ed.), *Vygotsky and education: Instructional implications and applications of sociohistorical psychology* (pp. 223–250). Cambridge: Cambridge University Press.

Hillocks, G., Jr. (1986). *Research on written composition: New directions for Teaching*. Chicago: ERIC Clearinghouse on Reading and Communication Skills: National Institute of Education.

Hillocks, G., Jr. (1995). *Teaching writing as reflective practice*. New York: Teachers College.

Hillocks, G., Jr. (2002). *The testing trap: How state writing assessments control learning*. New York: Teachers College.

Hillocks, G., Jr. (2003). Fighting back: Assessing the assessments. *English Journal, 92*(4), 63–70.

Kitagawa, M., & Kitagawa, C. (1987). *Making connections with writing: An expressive writing model in Japanese schools.* Portsmouth: Heinemann.

Kohn, A. (2004). Test today, privatize tomorrow: Using accountability to "reform" public schools to death. *Phi Delta Kappan, 85*(8), 568–577.

Moffett, J. (1988). *Coming on center* (2nd ed.). Portsmouth: Boynton/Cook.

National Writing Project & Nagin, C. (2003). *Because writing matters: Improving student writing in our schools.* San Francisco: Jossey-Bass.

NCAC. (2002) *Examples of literary works altered on New York State Regents English Language Arts Examinations.* Retrieved November 11, 2004, from http://www.ncac.org/education/related/20020611-NY-Albany-Examples_from_Regents.cfm.

NYSCLU. (2002) *Coalition of groups denounces censorship on Regents English Exam.* Retrieved November 11, 2004, from http://www.nyclu.org/news/coalition-of-groups-denounces-censorship-regents-english-exam.

NYSED. (1996) *Learning standards.* Retrieved November 11, 2004, from http://www.emsc.nysed.gov/ciai/ela/pub/elalearn.pdf

NYSED. (1999). *Description of SURR school groups.* Retrieved November 14, 2004, 2004, from http://www.emsc.nysed.gov/nyc/PDFs/SURRDescr.pdf

NYSED. (2004) *Information booklet for administering and scoring the Regents Comprehensive Examination in English.* Retrieved November 1, 2004, from http://emsc33.nysed.gov/osa/inform/DET%20541E%200604.pdf

Popham, W. J. (2001). *The truth about testing: An educator's call to action.* Alexandria: ASCD.

Rothman, R. (2001). One hundred fifty years of testing. In *The Jossey-Bass reader on school reform* (pp. 419–433). San Francisco: Jossey-Bass.

Smith, F. (1986). *Insult to intelligence: The bureaucratic invasions of our classrooms.* Portsmouth: Heinemann.

Tharp, R. G., & Gallimore, R. (1988). *Rousing minds to life: Teaching and learning in social context.* New York: Cambridge University Press.

Tucker, M. S., & Codding, J. B. (2001). Setting high standards for everyone. In *The Jossey-Bass reader on school reform* (pp. 470–482). San Francisco: Jossey-Bass.

USDE. (2003). *No Child Left Behind.* Retrieved November 13, 2004, from http://www.ed.gov/nclb.

|| SIX

The Corporate Model of Schooling

How High-Stakes Testing Dehumanizes Education

Morna McDermott

> Our schools are in a sense, factories in which raw materials are to be shaped and fashioned into products to meet the various demands of life. The specifications for manufacturing come from the demands of the 20th century civilization, and it is the business of the school to build its pupils to the specifications laid down. This demands good tools, specialization machinery, continuous measurement of production. (Cubberly, 1919, p. 503)

Introduction

Schools in the United States, at least since the industrial age, have been vehicles of social control. Factory model schools, designed during the industrial era and guided by the industrial paradigm, served that framework through economic, ideological, and political means. Now, just as decades ago, high-stakes testing (HST) is the weapon of choice used by education reformers to manipulate the educational system in ways that benefit their agenda to privatize public, pushing a standardized and highly regulated curriculum (to match with the required tests), increased social engineering (using and tracking student data via the HST for other purposes), and corporate profit (through the development, implementation, and evaluation of the HST). One cannot deeply understand the origins or purposes of today's high-stakes tests without examining the social, political, and economic climate in which they exist. High-stakes testing is the thread that ties together a larger picture of reform that includes privatization of public education, replacing public schools with charter schools, enforcing a curriculum that "force feeds" meaningless data to already disempowered and disenfranchised communities and uses "accountability" to turn data into big profits. Each of these issues, as they interface with testing policies and effects, will be explored in this chapter.

Education in the United States has always been overshadowed by elements of social engineering such as forcing English-only policies on indigenous groups and the perpetuation of Eurocentric "American" values as the core of the curriculum. Now we are a postindustrial society, and private control of means and ends of production by multinational corporations have replaced the factory model for our society. According to Q Finance (http://www.qfinance.com/dictionary/postindustrial-society), a postindustrial society is one

> [i]n which the resources of labor and capital are replaced by those of knowledge and information as the main sources of wealth creation. The postindustrial society involves a shift in focus from manufacturing industries to service industries and is enabled by technological advances.

Our approaches to schooling have followed suit, because now, like then, they continue to serve as a form of social control. The central vehicle used for control is testing. In the early 1900s scientific methods of testing human intelligence, for purposes of eugenics, were hailed as a mark of "social progress." Now, the reaches of social control through intelligence testing, for the benefit of a few at the expense of the masses, has reached unprecedented heights. Schools as a form of social manipulation have always been aimed directly at disenfranchised groups (i.e., people of color, people living in poverty, and immigrants). When we think of factory schools few people conjure images of white suburban schools attended by the Cleaver children. Factory schools were predominant in urban, poor, immigrant, and working-class communities.

Corporate models of reform (and the incessant push for more testing) dominate the educational landscape in disenfranchised communities still, where a quality public education was never given an opportunity to fully manifest itself. Deemed failures, schools in socioeconomically challenged neighborhoods, one by one, are being turned over to private educational entrepreneurs, who claim to have the magic pill to ease our social woes. They are not factories anymore; they are financial markets. The assembly line of streamlined efficiency of factory production as a model for the management of school learning has shape-shifted into something even more insidious. No Child Left Behind (NCLB) and Race to the Top (RtTT) have become the latest models of efficiency and productivity, and high-stakes testing is the vehicle used to achieve their goals. The centerpiece in the "schools as financial free market" are the big data sets: the ones corporations are paid to develop and assess—and the ones by which (disenfranchised) schools, teachers, and children are measured.

This chapter takes its reader not to the oft-cited analogy of schools as prisons as factories of industry. While these analogies are biting and accurate, this chapter instead explores how high-stakes testing promoted by current education reform policies are turning public schools into institutions that resemble corporate farms. These farms are owned by a handful of large corporations like Perdue, throwing their financial weight behind legislation that promotes their own economic and political interests, at the expense of small farmers, consumers, and the animals they keep. Like current education trends, where educational policy is increasingly owned by a handful of powerful corporations and lobby interests, large corporate-run farms, commonly called concentrated animal feeding operations (CAFOs), bring a host of complex social, political, and economic concerns with them.

Schools as CAFOs

In both institutions—CAFOs and schooling—there exist a nearly identical social, economic, and ideological paradigm shift; in this paradigm agriculture as a corporate-owned institution is now being imitated by the institution of schooling, and with similar results. This paradigm shift occurs in a few steps: (1) a crisis in the institution is maximized (or perhaps even manufactured), (2) which demands that new regulations be implemented, and then (3) the larger corporations collude with government agencies to craft legislation to re-form that institution, thus (4) creating openings for large corporations to move in and to push out small farms (in agriculture) and community public schools and community-run charter schools (in education). To make these steps manifest, data sets are needed to provide "scientific proof" of the crisis. Additionally, reformers make claims that more oversight and social management (via more—new and improved—testing) is needed to "solve" the crisis.

While we do indeed face serious problems with both hunger/food production and quality public education in this country, the solution to neither is to hand the modes of production over to major corporations spinning the roulette wheel of the free market. We cannot feed starving children tests. In both instances, institutional regulations are manipulated to favor large corporate interests, which have spelled trouble for communities and consumers, powerless in the face of this corporate/government power bloc. In both instances, large corporations turn enormous profits, while safety or accountability legislation does little to protect the persons or communities they were supposedly designed to serve. According to Falconer (2010), "though animal agribusiness giants have long touted their ability to produce cheap food for the masses, supermarket prices largely mask a larger hidden reality: Taxpayers shell out billions to prop up an inhumane, inefficient, and environmentally destructive industry" (p. 10).

The comparisons between these two entities are not merely flowery yet grim analogies. The paradigm of capitalistic production driven by the interests of a handful of corporate powers in food production is the same paradigm under which corporate reform of education operates schools. Therefore, schools are shaped not *like* CAFOs, but they *are* CAFOs. Like animals on CAFOs, schools and children are assaulted simultaneously by an overstimulation and under-stimulation of the senses. Children in schools are not even valued as consumers anymore, but as capital: raw material to be manufactured and processed as a product, treated not unlike animals in CAFOs.

In the wake of NCLB, RtTT, the uses and abuses of high-stakes testing have been detrimental to all forms of sensory development such as aesthetic education, spatial awareness, and tactile interactions with the world around them. We dull and deafen the senses of our students. In an effort to produce maximum capacity we hold them down at desks like cows in feedlots or chickens in restraints and pour pre-digested feed down their throats. In CAFOs, the quality of the food is minimal. Force-feeding corn and soy in animals who are naturally grass eaters mirrors the required use of one-size-fits-all standardized content in education, which is of little nutritional (meaningful or useful) value. The goal is merely to fatten students up to take their roles in the pecking order (no pun intended) in the capitalist/consumer world. Rather than asking how tethering students to high-stakes tests is *like* tethering chickens to battery cages with no daylight and fresh air, force-feeding them corn feed until they are unable to move, we should be asking ourselves, how are the two *not* different?

Both industries are dedicated to what Foucault called "efficient regimentation of docile bodies" (in Cole, 2011). According to Novek (in Cole):

> Notable institutional sites for the emergence of disciplinary power in nineteenth century Western society included prisons, hospitals, schools and barracks, wherein inmates were variously subject to correction, cure, education and drill in relation to norms of lawfulness, health, employability and obedience to military authority. (p. 86)

Cole, who is writing about animal farming, continues that "docility, argued Foucault, depends on the distribution of individuals in space such that they are amenable to surveillance and the production of individualised knowledge" (p. 86). *Surveillance techniques* such as tracking student data from pre-K to college, and the use of microchips in students uniforms, and the *production of individual knowledge* controlled by the Common Core State Standards (CCSS) curriculum and standardized tests, reveals how closely the paradigms of CAFOs and education reform are aligned.

Children are the conduit for actual products: money and data. Children's physical (docile) bodies are necessary to collect the dollars, through per-pupil funding and through the purchasing of corporate-sponsored materials such as tests, online services, and textbooks. And, students are needed to produce the data that must be evaluated and processed (courtesy of for-profit corporations) at an additional cost to society. In the words of Sir Ken Robinson (2006), we "mine children's minds the way we mine the earth for natural resources." They have been stripped of the tools needed to provoke change—the first of which is to feel and care about making change.

Is such conformity and standardization in CAFOs actually making the food we eat safer and more healthful? And are these measures in schools actually improving quality of education as being accessible, equitable, or meaningful? I contend that the answer to both is a resounding no. The system, constructed by the state in direct fealty to corporations, creates a complex opaque tangle of new regulations aimed at maximized social control (called accountability) and deregulations for maximized corporate profits (called *school choice*).

I offer this comparison not by simply applying one or two broad strokes in colorful comparisons, but by including fine brushstrokes through the juxtaposition of quotes or facts from the corporate agribusiness with facts and data pertaining to high-stakes testing and school "reform." This chapter addresses comparisons across five categories: high-stakes testing, teaching, curriculum, accountability, and privatization. Education, like the food industry, faces "the ascendancy of market economy power (which) has resulted in many localized relationships of hegemony as well as the wide-ranging hegemony known as 'globalization'" (Bosenberg, 2003).

High-Stakes Testing

Because of high-stakes testing (HST), beginning with the scare from *A Nation at Risk* (National Commission on Excellence in Education, 1983), morphing a decade later into NCLB, and making its crescendo with RtTT, the power of social engineering has reached new highs. These trends, built on a climate of fear, thus demand the need for school, student, and teacher surveillance. Teachers know they're being watched from above more now than ever. Their every move is micromanaged, supervised, and tethered to tests that provide facts completely disassociated from context—children regurgitate knowledge devoid of personal meaning. In states such as New York, individual teacher test scores are published in public venues. The purpose is to reinforce the hegemonic gaze of those who wish to engineer and control the every move of not only teachers but also students. Tests are now being taken on computers where test scores are immediately logged into a database somewhere, and kept by a third party. We no longer even own the property rights to what we know. HST is about social control or management: an efficiency model for managing large numbers of people. It is not about providing the best quality education to all children.

Table 6.1. Animals as Food-producing Machines Compared to Students as Data-producing Machines.

CAFO Facts	**Facts about High-Stakes Testing**
"Because factory farms are considered 'agricultural' instead of industrial they are not subject to the regulation that their scale of production warrants" (Factory Farming Facts).	Because testing is mandated under regulations of school "accountability" rather than being evaluated for its pedagogical soundness, Pearson uses children as testing guinea pigs for creating their future tests (see New York City Teachers, 2012); the testing industry operates with little regulation other than those written to serve its own needs.
The use of confined high-pressure environments to produce maximum yield and profit. Chickens' beaks are routinely cut off to prevent them from pecking one another. Molting and other skin conditions due to stress are prevalent and treated with more antibiotics.	Children have been known to begin scratching themselves nervously and developing anxiety-related ailments because of the testing conditions not unlike animals that begin biting or licking at their own skin out of anxiety or boredom, only to become deathly ill and diseased as a result.
Animals are confined to conditions where they are forced to live in their own waste. According to *Farm Sanctuary* (www.farmsanctuary.org), "In factory farms they are forced to live in their own feces."	In one recent instant, a kindergartner was forced to sit in soiled pants during a school test: "A six year old told her teacher she had to go to the bathroom during a test and wasn't allowed . . . she still had poop, diarrhea poop, coming out the back, up her front, and down her legs . . . . All the while, the little girl had to sit in the mess" (Forhetz, 2012). There is another recent report of a seventh-grade student in Texas who was forced to urinate in a bottle after the teacher denied him a bathroom break during testing (Hibbard, 2012). And what happened next? He faced one month in "an alternative campus" as punishment.
In the chicken industry, enormous quantities of food are thrust down their throats through a tube.	In schools, test-driven knowledge is forced down children's throats, fattening them up for test-taking day, when such information can be extracted for profit.

Data for the last 10 years (Amrein & Berliner, 2002) indicate that high-stakes testing has had an overwhelmingly negative effect on the same populations that they claim to serve. But HST continues en masse at greater speed and with increased frequency. Students are no longer considered learners; they are data-producing machines to serve the Chamber of Commerce, the military industrial complex, and testing companies who promise (for a small fee) to bring solutions to the same problems they helped to create in the first place. While public schools continue to close, and students continue to drop out or perform even more poorly than they did 10 years ago, Pearson has made billions of dollars from the testing industry. Just as animals are not considered animals at all—they are food-producing machines—so our students have become data-producing machines (see Table 6.1).

Teachers

In the teaching profession, two phenomena are underfoot, spurred on by new uses of technology: increased surveillance and deskilling the teaching labor force. New teacher education programs are required by new policies to ensure that student teachers demonstrate successful K–12 student outcomes, or else not graduate. CCSS now shape the Teacher Preparation Assessment (TPA) in teacher preparation colleges (both contracted with Pearson). Preservice teacher data too will be monitored and tracked. Teacher educators will be held responsible for the potential failure of future teachers before they've even walked across the stage. The great panopticon (Foucault, 1977) will be watching us all and controlling our every move.

With the current push to increase class size, and to eliminate teaching positions in favor of online classes, electronic assessment of tests, and educational services such as Common Core lesson plans prepackaged for download (all courtesy of Bill Gates and Pearson), the teaching profession is becoming a deskilled labor force. The effects are many, but one central concern is the increasing rift between those in power and those without voices. As Kesson (2004) illustrates:

> White elite voices speak from one face about fixing failing schools, providing equity and access, raising the standards for all, and leaving no child behind, the muted voice of the hidden face reminds us that is the nature of the modern administrative state to disperse power (accountability) in accordance with a mode of rationality which dictates that any such dispersal must ultimately result in a strengthening of that power. (p. 37)

Such deskilling of the educational profession for teachers in disenfranchised schools "is not merely a professional issue . . . it is a political issue, as it is employed not just to control the labor of teachers, but also to ensure conformity of student thinking" (Kesson, 2004, p. 32). More and more, technology is being used to further deskill teachers and to maximize production, thus further alienating students and teachers from their own experience. Computers will replace real humans as delivery models. Skilled farm labor has been increasingly replaced with machines that can perform the same job, cheaper and more efficiently but with less humanity as well. In comparison:

> [T]he coming together of the computer revolution and the biotechnology revolution into a single technological complex foreshadows a new era of food production—one divorced from land, climate and changing seasons, long the conditioning agents of agricultural output. In the coming half century, traditional agriculture is likely to wane, a victim of technological forces that are fast replacing outdoor farming with manipulation of molecules in the laboratory. (Rifkin, n.d.)

Table 6.2. Technology in Farm Labor and Teaching

Technology and Farm Labor	**Technology and the Teaching Profession**
Robots are being developed with artificial intelligence to plow and seed fields, feed dairy cows, even shear live sheep. Researchers predict that the fully automated factory farm is less than 20 years away (Rifkin, n.d.).	Gates and company will have students tethered to computers inputting their data for someone else's purpose rather than allowing schools to provide hands-on interactive and creative learning experiences.
According to the USDA (http://www.ers.usda.gov/amber-waves.aspx) extraordinary increases in agricultural labor productivity due to technological innovation reduced farm employment from almost 10 million in 1950 to just over 3 million by 2006. At the same time, agricultural output increased by more than 150%; over time, agricultural production has concentrated on fewer and larger farms.	Bill Gates is buying the education system in his own vision with increased class sizes, elimination of tenure for teachers, and union busting strategies (deVise, 2011). As class sizes increase there is a need for technology via online courses and computer-based learning and assessment to fill the void teachers will leave behind.
The primary goal of industrial farms is to maximize profits—even if it threatens the well-being of farmworkers, the men and women who help bring food to our tables. Workers on industrial farms and those in the food-processing industry are often subject to hazardous working conditions and unfair labor management practices.	In CAFO-like schools (largely poor and urban with students and teachers of color), "minority children are being taught by teachers who work under oppressive conditions, subject to a tyrannical management paradigm that stifles their professional growth and thus undermines genuine and long lasting improvement of their schools" (Kesson, 2004, p. 36).

continued on next page

Technology and Farm Labor	Technology and the Teaching Profession
The decline in the number of farms is likely to accelerate in the coming years with advances in agricultural software and robotics that will lead to higher yields and fewer workers. A new generation of sophisticated computer-driven robots may soon replace many of the remaining tasks on the land, potentially transforming the modern farm into an automated outdoor factory.	Eliminate or reduce "seat time" requirements for students to be with licensed staff, focusing on student outcomes (aka tests) instead. This will supposedly "allow, for example, unlicensed staff to monitor digital labs, freeing funds to pay more to the excellent teachers in charge" (Hassell & Hassell, 2011, p. 9).
Tractors that know exactly where to plow and sow seed, as well as turn corners, are soon to hit the market. Preprogrammed carts that follow harvesting combines, and automatically steer themselves to waiting trucks when they get full, are also in the works. And the utilization of advanced new software, censors, and GPS technologies is set to switch management of much of the rest of the food-growing process from humans to machines as well (Benson, 2011).	The Common Core Online Curriculum Toolkit provides resources that will assist educators in designing instructional programs that are aligned with the new curriculum and the new assessments. The toolkit will contain model lessons, model units, formative assessments, multimedia resources, intervention and enrichment modules, and online courses for students and educators.

In the realm of new technological advances, teachers become proctors, or distance-learning "speech" givers. What is lost is the immediate and sensory-based (McDermott, 2011), interactive, site-based, and hands-on learning within a community of learners. Just like the chilling account in *The Education of Sam Sanders* (Poetter, 2006):

> The scores go directly to the state computer; the state closely monitored each test site by computer and by camera. He could not coach the students in any way, shape, or form; he couldn't give them any sorts of cues or prompts. . . . All the while he knew the computer wouldn't tabulate a score for Sam, and that Sam's "0" would count against the class, and against him. (p. 41)

More comparisons are listed in Table 6.2.

Curriculum

For the purposes of this chapter, I include various facets of curriculum beyond just the material being delivered, to include curriculum as space (where we teach) and pedagogy (how we teach). Children serve as actual products when we place them in CAFO models of schooling where they are little more than trained to become mindless livestock, fed on a strict diet/curriculum of basic skills dictated by businesses, to be molded into the products (as human capital) corporations desire to serve their own outcomes. It's no wonder that the U.S. Chamber of Commerce is vying to have a seat at the table of school boards, hand-picked and appointed, so they can manipulate local educational policies to their liking. Large corporations have now subsumed education as one of their markets.

Like large CAFO farms that place profit over any concerns for human or animal rights, corporations shaping educational policy have less concern for children and communities than they do for efficiency and finances. Inhumane conditions exist all over the country for students forced to attend schools that have little regard for their human welfare, but enormous regard for the policymakers who use children as a form of profit.

Curriculum has become like corn feed and antibiotics. It is designed for maximum production, not the nutritional quality that's best for the child, but what increases production and profits. In China, Huang (2012) published a story about a school where students are given amino acids intravenously while taking tests to boost their testing performance. Arts-based learning, critical thinking, exploratory learning, small inquiry group—all take time and money, none of which are beneficial to "production." A straight diet of testing materials offers the most direct and efficient means for producing test results.

Another way that CAFO farms increase productivity and reduce costs is by manipulating the diets of the animals, not feeding them what might be nutritionally sound or even humane, but what will fatten them up the fastest and cheapest. Animals that normally and naturally would feed on grasses and bugs are fed only corn feed. The Common Core State Standards are revealing themselves to be little more than corn feed. Despite rhetoric, a real examination of the CCSS reveals the intellectual range of dietary options for children's minds, and imaginations are severely limited. A few comparisons are made in Table 6.3.

Table 6.3. Schools as CAFOs

CAFO Facts	**Schools as CAFOs**
In CAFOs unnaturally large numbers of animals are confined closely together.	Increased class size and forced passivity of extended seat work create behavioral and management problems.
Animals are confined to metal buildings, indoors, with minimal room for normal behaviors and little or no room for sunlight and fresh air.	Recess, art, music, and PE are eliminated in favor of more seated test preparation.

continued on next page

CAFO Facts	Schools as CAFOs
Low doses of antibiotics are administered regularly and animals are fed antibiotics to promote faster growth. With the approval of the FDA and USDA, CAFOs use hormones to promote growth and milk production.	Increased numbers of children are placed on Ritalin and other medications to increase compliance with overregulated schooling conditions and sedentary activities such as testing. I also refer again to the article about students in China receiving amino acids through IVs during test-taking for college entrance exams.
According to *Sustainable Table* (www.sustainable.org), "One corporation often owns and controls all aspects of the production process, including animal rearing, feed production, slaughter, packaging and distribution."	The house of education reform is lorded over by Bill Gates, Eli Broad, The Walton Foundation, and Pearson Publishing. Pearson alone now controls the GED, the SAT, the Common Core, much of the statewide high-stakes testing contracts, and Teacher Performance Assessment (TPA) for universities across the country.
The conditions in chicken houses for both broiler chickens and egg laying chickens are not only cramped and unsanitary but also include excessive heat and a lack of fresh air. They are confined to tiny, barren "battery cages" stacked down the aisles of windowless warehouses.	Students in low-income, urban and rural settings are routinely crammed into old buildings with no air conditioning or clean drinking water, forced to remain indoors and to perform test-oriented tasks at desks.

Accountability

State and federal legislation makes false claims that small farms are dangerous and therefore must be taken over by large corporations to save the food industry from disaster. They hide or bury facts that the large corporate-run farms that replace the small family farms, like the new corporate-run charter schools, perform no better than the community public schools (or family farms) they replaced. *GradNation* is one such organization that makes loud claims to promoting greater student regulation in the name of national security and improved graduation rates.

But look at who is at the *GradNation* table: State Farm Insurance Company, the Simon Foundation for Education and Housing, ING Foundation, Walton (Walmart) Foundation, AT&T, Boeing Company, Pearson Foundation, Jim and Donna Barksdale, DeVry, The Packard Foundation, Target Corporation, Philip Morris USA, an Altria Company, Fidelity Investments, Ritz-Carlton, Annie E. Casey Foundation, Rockefeller Foundation, and Wellspring.

Accountability has become the buzzword to sell their agenda to the public, but what it really spells is the opportunity for corporate-led interests to dominate education reform. Reformers both for and against the Common Core State Standards manipulate the language of policies for accountability to edge out public community schools and small community-based charter schools in favor of for-profit charter school chains, like McDonald's popping up in poor, disenfranchised communities. The American Legislation Exchange Commission, though it originally included Common Core State Standards as part of its "model legislation," recently backtracked on its endorsement. However, the goal is still the same—corporate profits. Accountability is about surveillance and profit. Proponents of the Common Core decry that standardized measures are necessary for accountability purposes, while reformer-opponents (not to be confused with a

Table 6.4. Farm Health Regulations and Education Accountability

Facts about Farm Health Regulation	**Facts about Accountability**
In the agriculture business, Salatin (2009) points out "the second argument is about food safety. 'How can we be sure that food produced on local farms without centralized inspection and processing is really safe to eat?'" (p. 185).	As Mathison and Ross (2002) note, "Accountability has become the means of enforcement and control used by states and businesses." It is a form of social control that can restrict individual opportunities while imposing the authority of official knowledge.
"Agribusiness was using the food crisis to gain public acceptance of genetically modified crops by claiming the miracle plants could end world hunger" (Pringle, 2009, p. 67).	*A Nation at Risk* (National Commission on Excellence in Education, 1983) swept the nation up into a frenzied panic that greater education oversight and control would be needed to solve the education "crisis." Now, *EdNation* has the Civic Marshall Plan (Resmovitz, 2012). This state of crisis opens the floodgate for reform measures that favor "genetically modified" schools: online education and charter school chains making promises to end the drop-out crisis and other educational "ills."
CAFOs are highly standardized for efficiency. Monocultures of animals and feed crops are created to be highly unified through gene manipulation to help yielding consistent production every year. Website Belsandia (www.belsandia.com) states, "Less diversity and variety of agricultural products make management and regulation of food quality easier."	The CCSS are being implemented in most states, creating a mono-curriculum. Less and less diversity of thinking can be seen as testing mandates eliminate creative student-centered classrooms. Less diversity makes regulation easier. Standardized tests can be evaluated en masse electronically, so all other forms of teaching and learning that are not efficient are eliminated.

continued on next page

Facts about Farm Health Regulation	Facts about Accountability
In the last few decades, consolidation of food production has concentrated power in the hands of fewer and fewer corporations. Many of today's farms are actually large industrial facilities, not family farmers with green pastures and red barns that most Americans imagine.	Charter schools are no longer individual community-based operations. Today's charter schools are largely the property of hedge fund venture capitalists and spread like fast food chains moving into communities pretending to be "grass roots" efforts.
The National Animal Identification System (NAIS) is a government-run program in the United States USDA (see http://www.aphis.usda.gov/traceability/) intended to extend government animal health surveillance by identifying and tracking specific animals. Administered at the federal level by the Animal and Plant Health Inspection Service, a branch of the United States Department of Agriculture, NAIS will also be overseen by state animal health boards. Money received by some states has been used to make parts, or all, of the program mandatory.	In consultation with the U.S. Department of Education, GradNation (sponsored by America's Promise Alliance) will "track ten measures that research has proven predict student success. We will also encourage communities to measure other factors. For instance, tracking school attendance, behavior indicators and course grades in math and English at the local level serve as a valuable early warning signal, even though reliable sources of data do not yet exist on the national level" (http://www.americaspromise.org/our-work/Grad-Nation.aspx).
According to *The True Food Network* (www.truefoodnow.org), the American Public Health Association has called for a moratorium on new factory farms because of the devastating effects these operations can have on surrounding communities. These consolidated operations have little to no regard for the environment, animal welfare, or food safety, and they often put the health of consumers and rural communities at risk for the sake of profit.	The NAACP has filed complaints against charter schools because of the devastating results of racial desegregation left in their wake. Children of color and children living in poverty are disproportionately being isolated in (or away from) charter schools that are without racial or socioeconomic diversity. Corporate model charter schools drain public community schools of funding and destroy community solidarity, as well as sapping them of resources and "voice."

real critique of the CCSS's association with corporate reform) claim CCSS promotes too much governmental oversight on state-led matters of schooling (see Table 6.4).

Privatization

In both institutions one can see the fingerprints of corporate interest in the guise of the American Legislative Exchange Commission (ALEC), behind the scenes, crafting model legislation that serves the interests of corporations aided and abetted by governmental bodies.

The Chambers of Commerce across the country are rolling over school boards, claiming to attempt to streamline decision making and promote a strong business partnership, and such actions create corporate oversight and decision making power that favors large corporate interests. Community members who resist these efforts are met with silence.

According to Bosenberg (2003):

> The privatization of public schooling is cause for considerable concern. Students and their families increasingly face corporate, for-profit schools with unproven credentials as their only alternative to currently existing public schools. Teachers face a weakening of their labor rights when for-profit schools hire non-union teachers in order to keep their expenses to a minimum. Education researchers risk marginalization of pedagogical innovations that may adversely affect a company's bottom line and condemnation of research that critiques private sector involvement in schooling. Social integrity and personal freedom are in jeopardy as corporations increasingly exert their influence on the curriculum and the discourse of schooling.

EdNation and similar corporate-minded panels and think tanks dominate the education policy landscape. They claim to provide nonpartisan interests dedicated to the welfare of children; their true interests lie with stockholder profits and ideological hegemony.

Like ALEC's model educational legislation being forced through states where for-profit corporate charters reign supreme, "the power wielded in Washington by lobbyists from the farm states remains enormous. And within the subsidy-rich world of U.S. agriculture, corn is king . . . with massive federal support and mandated production levels, bringing it to a halt is difficult" (Bryce, 2009, p. 92). The agenda for corporate model reform has been in the works since *A Nation at Risk* (National Commission on Excellence in Education, 1983).

The development of CAFOs, similarly, "is not sustainable, nor it is accidental. In large measure, it can be traced back to government policies designed to produce the very system that now distorts agricultural production in this country" (Rodrigues, Delwiche, & Kasooji, 2009, p. 129; see Table 6.5).

There Is Hope

In both institutions one can see the following in action: "the power of corporations to influence government policy . . . companies seeking deregulation—and government subsidies. They hate government regulations that protect workers and consumers but love to receive tax payer money" (Schlosser, 2009, p. 9).

Table 6.5. Privatization in Agriculture and Education

Privatization and New Agricultural Policies	**Privatization and New Reform Policies**
As Benson (2011) writes: "Globalization combined with ever-evolving government agriculture policies that favor factory farming have together helped create this dire and unsustainable system, which has left farmers with little choice other than to continually adopt whatever new methods and techniques come on the scene to quicken the growing process and boost yields, even if it means removing the human element almost completely."	In these times of dire economic recession most states and local districts have been given by RtTT "an offer they can't refuse." To survive, states feel compelled to comply with the demands of RtTT, adopting whatever new testing and techniques will be required, such as more funding for corporate charter schools, school vouchers, increased testing, and mandated curriculum.
"Of the many outrages of the corn ethanol program," according to Bryce (2009), "surely the most shameful aspect is this: Congress has mandated the creation of—and the payment of subsidies to—a multibillion dollar network of distilleries that are burning food to make motor fuel at a time where there is a growing global shortage of food and no shortage of motor fuel" (p. 92).	In spite of the failure of charter schools to evidence any real success, taxpayer subsidies in the form of vouchers are being poured into them, replacing their public school counterparts, though they rarely even outperform the schools they are replacing.
Political pressure, the power of giant agribusinesses, and willful ignorance of basic science, states Pringle (2009), "can be used to distort U.S. farm policy, producing results that include not only economic damage to the nation but the potential for increasing hunger around the world, especially among the most vulnerable people on our planet" (p. 66).	"The pro-voucher funding stream appears unstoppable, with sources like the Bill and Melinda Gates Foundation, the Eli and Edythe Broad Foundation and the Walton Family Foundation." Denvir (2012) illustrates how "the same political forces that have bled Philly schools for decades now decry their poor performance. The solution, of course, is the private sector."

continued on next page

aged, publicly owned, and publicly attended schools where real learning take place and high-stakes testing has been eliminated. Data-driven practices would be replaced with problem-based learning, the arts, child-centered pedagogy, fully funded resources, smaller class sizes, and a deep appreciation for the imagination and social justice. We can re-sensitize and re-humanize learning. As parents, we can boycott testing the way we boycott products in a grocery store.

Privatization and New Agricultural Policies	Privatization and New Reform Policies
Hettinger and Smith (2010) state that "last year, the industry spent millions trying to preempt animal welfare reforms through the creation of its own entity, the Ohio Livestock Care Standards Board" (p. 21). According to one activist, "I think the best you can hope for out of them is a 20-year study on battery cages."	Education "think tanks" overwhelm and confuse citizens and educators alike with bogus research such as how to align the Common Core with differentiation strategies, or how to improve testing to measure critical thinking skills that distract us from the real facts that we already know, and mire us in endless debates of no consequence.
According to Falconer (2010), "For years the federal government has used public funds to subsidize grain farmers whose production expenses exceed market rates . . . taxpayers have also subsidized remediation efforts when waste containment systems fail" (p. 10).	"After being deemed failures," Bosenberg (2003) explains, "public schools taken over by for-profit corporations are no longer completely accountable to the public yet are still paid for with taxpayers' money. Recent legislation that employs the language of *choice*, *efficiency*, *effectiveness*, and *accountability* will greatly facilitate the process of publicly-funded school privatization."

How do we combat high-stakes testing-driven CAFO models of schooling? We, as educators, can take some advice from the advocates for more humane and sustainable farming practices. Maybe we could start envisioning free-range schools with grass-fed learning. These would be publicly funded, publicly man-

The arts/aesthetics in themselves do not change anything, but when centered within and around community and social concerns, aesthetic experiences hold the possibility for change by reawakening (Greene, 1995) the senses; pulling us from the torpor—the feel, to sense, to participate—in that possibility we can hold a bit of hope, and hope is the solution to despair. Rather than continuing to allow millions of children to be tested for the purposes of social engineering or as products to be used by the elite for elitist agenda, we can educate students to make the world "as if it might be otherwise" (Greene, p. 47).

Salatin (2009) calls for a "Food Emancipation Proclamation." I call for an "Education Emancipation Proclamation"—a demand to be freed from the tyranny of corporate greed and self-interest. I suggest we explore ways to reclaim child-oriented pedagogy, innate human desire to play, to be curious, to want to learn, to embrace organically (no pun intended) emerging teaching and learning processes in schools much in the way that new sustainable farming practices are grounded first in embracing the natural instincts of animals and honoring the earth. Like Salatin (2009), to begin "anew," we must first opt out of the corporate order: "At the risk of being labeled simplistic, I suggest the most efficacious way to change things is simply declare our independence from the figurative kings of the industrial system" (p. 188).

References

Amrein, A. L., & Berliner, D. C. (2002). High stakes testing, uncertainty, and student learning. *Education Policy Analysis, 10*(18). Retrieved from http://epaa.asu.edu/epaa/v10n18/

Belsandia. (n.d.). *Factory farming facts—Animals raised for food cause human health risks, endanger the environment, and endure animal cruelty*. Retrieved from http://www.belsandia.com/factory-farming.html

Benson, J. (2011). The dehumanization of agriculture: New robotic farming devices eliminate need for human farmers. *Natural News*. Retrieved from http://www.naturalnews.com/033701_agriculture_robotics.html

Bosenberg, E. (2003). *Privatizing public schools: Education in the marketplace*. Retrieved from http://louisville.edu/journal/workplace/issue5p2/boesenberg.html

Bryce, R. (2009). The ethanol scam: Burning food to make motor fuel. In K. Weber (Ed.), *Food Inc.* (pp. 91–103). New York: Public Affairs.

Center for Food Safety. (n.d.). *Factory farms*. Retrieved from http://truefoodnow.org/campaigns/factory-farming/

Cole, M. (2011). From "animal machines" to "happy meat"? Foucault's ideas of disciplinary and pastoral power applied to "animal-centered" welfare discourse. *Animals, 1*, 83–101.

Cubberly, E. (1919). *Public education in the United States*. Boston, MA: Houghton Mifflin.

deVise, D. (2011). *Bill Gates talks about teacher pay, and class size*. Retrieved from http://voices.washingtonpost.com/collegeinc/2011/02/bill_gates_talks_about_teacher.html

Denvir, D. (2012). Who's killing Philly Public Schools? Underfunded. Overburdened. About to be sold for scrap. *Philadelphia City Paper*. Retrieved from http://www.citypaper.net/cover_story/2012-05-03-whos-killing-philly-public-schools.html?viewAll=y

Factory Farming Facts. (n.d.). Retrieved from www.idausa.org/facts/factoryfarmfacts.html

Farm Sanctuary. *Factory farming*. Retrieved from www.farmsancutuary.org

Falconer, J. (2010, July/August). The illusion of cheap food. *All Animals*, pp. 10–11.

Forhetz, S. (2012, April 17). Test turns to humiliation for kindergartner: Six-year-old was not allowed to go to restroom. *KY3 News*. Retrieved from http://articles.ky3.com/2012-04-17/missouri-assessment-program_31357428

Foucault, M. (1977). *Discipline and punish: The birth of the prison*. New York, NY: Vintage.

Greene, M. (1995). *Releasing the imagination: Essays on education, the arts, and social change*. San Francisco, CA: Jossey-Bass.

Hassel, E. A., & Hassel, B. C. (2011). *Seizing the opportunity to the top: How the U.S. can reach every student with an excellent teacher* [Policy brief]. Retrieved from http://opportunityculture.org/seizing_opportunity_policybrief-public_impact.pdf

Hettinger, J., & Smith, K. (2010, July/August). All eyes on Ohio. *All Animals*, pp. 21–22.

Hibbard, L. (2012). *Missouri teacher forced 6-year-old to sit in her diarrhea*. Retrieved from http://www.huffingtonpost.com/2012/04/18/missouri-teacher-forced-6-year-old-to-sit-in-her-diarrhea_n_1435592.html

Huang, J. (2012). Getting amino acids? Student receiving IV drips during study. *China Hush*. Retrieved from http://www.chinahush.com/2012/05/08/getting-amino-acids-student-receiving-iv-drips-during-study/

In Defense of Animals. (n.d.). *Factory farm facts*. Retrieved from www.idausa.org/facts/factoryfarmfacts.html

Kesson, K. (2004). Inhuman powers and terrible things: The theory and practice of alienated labor in urban schools. *Journal for Critical Education Policy Studies (2)* 1, 31–47.

Kirby, D. (2010, July/August). From homeland to wasteland: How I learned firsthand about the assault on rural America. *All Animals*, pp. 15–20.

Mathison, S., & Ross, E. W. (2002). The hegemony of accountability in schools and universities. *Workplace.* Retrieved from http://www.louisville.edu/jounral/workplace/issue5p1/mathison.html

McDermott, M. (2011). Curriculum of the eye/I. *Journal of Curriculum Theorizing, 27*(2), 130–144.

National Commission on Excellence in Education. (1983). *A nation at risk.* Washington, DC: USA Research.

New York City Teachers. (2012, May 29). *NYC teachers support parents on field test boycott in letter to Walcott.* Retrieved from http://ednotesonline.blogspot.com/2012/05/nyc-teachers-support-parents-on-field.html

Poetter, T. S. (2006). *The education of Sam Sanders.* Falls Village, CT: Hamilton.

Q Finance. (n.d.). *Definition of post industrial society.* Retrieved from http://www.qfinance.com/dictionary/postindustrial-society

Pringle, P. (2009). Food, science, and the challenge of world hunger-Who will control the future? In K. Weber (Ed.), *Food Inc.* (pp. 61–65). New York: Public Affairs.

Resmovitz, J. (2012). School report: Failing to prepare students hurts national security, prosperity. *The Huffington Post.* Retrieved from http://www.huffingtonpost.com/2012/03/19/schools-report-condoleezza-rice-joel-klein_n_1365144.html

Rifkin, J. (n.d.). *New technology and the end of jobs.* Retrieved from http://www.converge.org.nz/pirm/nutech.htm

Robinson, K. (2006). *Do schools kill creativity?* Retrieved from http://www.ted.com/talks/ken_robinson_says_schools_kill_creativity.html

Rodrigues, A., Delwiche A., & Kasooji, S. (2009). Cheap food: Workers pay the price. In K. Weber (Ed.), *Food Inc.* (pp. 123–143). New York: Public Affairs.

Salatin, J. (2009). Declare your independence. In K. Weber (Ed.), *Food Inc.* (pp. 183–197). New York: Public Affairs.

Scholosser, E. (2009). A conversation with Eric Schlosser. In K. Weber (Ed.), *Food Inc.* (pp. 3–19). New York: Public Affairs.

Sustainable Table. (n.d.). *Issue: The workers.* Retrieved from http://www.sustainabletable.org/issues/workers/

True Food Network. (n.d.). *Factory farms.* Retrieved from http://truefoodnow.org/campaigns/factory-farming/

U.S. Department of Agriculture. (n.d.). *Animal disease traceability.* Retrieved from http://www.aphis.usda.gov/traceability/

|| SEVEN

Standardized Testing and Boredom at an Urban Middle School

Richard Mora

The growing import given to high-stakes tests brings to mind the state-run light bulb factories that existed in the former Soviet Union. Those under-resourced factories were required to meet mandatory monthly quotas, not of the number of bulbs but of the total combined bulb wattage produced. So, in the last days of each month, if factory managers found themselves behind schedule, many of them would instruct their workers to produce larger bulbs of higher wattage. Though the factories filled up with outrageously oversized bulbs that were unusable, as the country had no sockets large enough to accommodate them, the managers' problem was solved—quotas were met. Like the Russian factory workers of yesteryear, present-day public school teachers in the United States are required to meet state-imposed mandates that are also resulting in absurdly problematic consequences. They are charged with using specific curricula to produce students who can excel on standardized tests, while the politicians and policy entrepreneurs promoting these educational reforms give no mind to how their efforts are dulling and boring young learners.

Decades of federal mandates for greater educational accountability and the subsequent "pervasive testing culture" (Moses & Nanna, 2007, p. 55) have drastically altered public education. In many classrooms more time is now spent on test preparation, at the expense of engaging and varied learning activities (Abrams, Pedulla, & Madaus, 2003; Shepard & Dougherty, 1991; Smith, Edelsky, Draper, Rottenberg, & Cherland, 1990). There is also substantial evidence of curricular narrowing to the subject matter on tests and an increase in "the use of teacher-centered pedagogies" (Au, 2007, p. 258). Some teachers report that in order to focus on testing, they discontinued

> instructional activities that (a) are pleasant for the teacher and the children; (b) provide reinforcement of skills and promote in-depth understandings of content; (c) involve collaboration, independence, and higher order thinking skills; and (d) have goals that are not measured by tests (such as the development of attitudes). (Barksdale-Ladd & Thomas, 2000, p. 391)

All these changes to public education and others have come about even though high-stakes exams result in little, if any, increased learning among students (Amrein & Berliner, 2002; Klein, Hamilton, McCaffrey, & Stetcher, 2000; Koretz, Mitchell, & Stetcher, 1996). What is more, as this chapter shows, the testing culture is creating un-stimulating learning environments, regularly exposing students to a major factor in academic disengagement and dropping out—boredom.

"School Is *So* Boring"

While conducting a multiyear, gender study at an urban K–8 school, I witnessed and documented the ground-level impact the push toward greater accountability in public education had on the group of 33 working-class Latina/o students that I followed. At Romero, as I call the school, standardized test scores served as the ultimate measure of the school's performance. As a result, entire class periods, hours at time, were dedicated to both district and statewide assessments, with teachers teaching to the test, to the practice tests, and to pre-practice tests. During these tests and the various quizzes and exams their teachers administered, the students had to sit quietly at their desks for long stretches of time, an expectation that proved difficult for most.

Additionally, during the sixth grade, the majority of students I observed had a double math period meant to prepare them for the upcoming state exam. Students found these experiences excruciatingly frustrating and repeatedly summed up their feelings with some variant of the statement, "School is *so* boring." On its face, the statement is not noteworthy. The phenomenon of boredom, which refers to situated experiences that are not entertaining and/or lack meaningfulness (Barbalet, 1999; Brissett & Snow, 1993), is associated with schooling (Anderson & Ridley, 1978; Beaulieu, 1981; Corrigan, 1979; Dow, 2007; Fallis & Opotow, 2003; Farrell, Peguero, Lindsey, & White, 1988; Fogelman, 1976; Healy, 1984; Heron, 2003; Keiler, 2011; Larson & Richards, 1991; McGiboney & Carter, 1988; Nelson, 1985; Robinson, 1975; Rubin, 2007; Wasson, 1981; Willis, 1977). Consequently, the complaint seems little more than just another flippant remark made in a public U.S. school.

However, an examination of exactly what it was that the Romero students deemed "boring" reveals telling criticisms of their schooling. My observations and conversations with the students strongly suggest that the students' boredom was largely due to a disconnect they felt between the act of learning and curricular practices focusing primarily on test preparation. In line with previous findings (Anderson & Ridley, 1978; Beaulieu, 1981; Corrigan, 1979; Larson & Richards, 1991; Willis, 1977), students' boredom at Romero was not correlated with academic ability or achievement. All Romero students expressed that they experienced some amount of boredom. They regularly complained that class instruction, particularly in their math and science classes, needed to be "more interesting." The students repeatedly expressed their desire for curricular experiences with more relevance, asking their teachers some variant of the question, "Why do we have to know this?" If, and when, teachers responded that students needed to know the curriculum presented because it would be covered in a test, students displayed their displeasure by shaking their heads, sighing loudly, and/or sucking their teeth. The students' apparent disgust was directed mostly at the various standardized tests they were prepped for and which school officials used to determine whether students should be promoted to the next grade level.

Like other students (Fallis & Opotow, 2003; Heron, 2003; Keiler, 2011), the students at Romero frequently grew bored with lecture-driven courses. During the sixth grade, the students bemoaned the double math period because it involved having to sit and listen to lectures from their teacher or silently work on math problems from their textbooks and test-preparation materials. In addition, the students found it particularly unfair that they were the only students in the middle school who had a double period of "boring math" instruction that rarely included group activities or projects. One afternoon, the boys in the class made their displeasure known. They rebelled by going "on strike." Here is how Leonardo, one of the boys, described the incident, while smiling:

> we didn't take out our math books. We came back from the park [late] with Coach Taylor and asked Ms. Connor if we could drink water and she said no. So, we weren't going to do math. None of the boys took out their books. We just talked.

While the impetus for the boys' strike was their perception that Ms. Connor was unreasonable, they were not seeking to engage her in a power struggle; far from it. Rather, they readily agreed among themselves to refrain from participating in math class that day because as strongly as they believed they deserved water, they were even more adamant that having to experience back-to-back "boring math" periods was the greater injustice.

Later that afternoon, Ms. Connor told the boys, "The good in me is dead. I'm not going to be nice anymore." She subsequently shared with me that she was frustrated because she tries hard "to teach [the boys] math" and prepare them for the state exam because many of them had not done well on the math section on their previous exam, and yet the boys refused to follow along while she explained the material on the chalkboard. It was not clear whether Ms. Connor knew that her lectures "bored" some of her students. Nor was it clear whether she knew that bored students are more apt to be disruptive (Fogelman, 1976; McGiboney & Carter, 1988; Robinson, 1975; Wasson, 1981) and that students having a hard time with the subject matter are more susceptible to boredom (Keiler, 2011).

Though the boys' behavior during math class may give the impression that they rejected schooling and its societal value, they did not. Quite the contrary, in fact. The boys, like their female classmates, actively engaged in class activities that, according to them, were "not boring" and gave them an opportunity to "talk" or "share." In their other sixth-grade classes, the boys regularly, enthusiastically raised their hands and volunteered when the teacher asked a question or asked for students to share their written work. In those classes, the boys also occasionally affirmed each other's academic success in expressive manners. The clearest example is a spontaneous celebration that erupted in social studies class after a geography quiz on ancient Indian civilizations:

> Immediately after each student turns in his or her quiz, Ms. Ross grades it, records the grade, and hands it back to the student. After receiving his grade, Carlos walks from Ms. Ross's desk over to me smiling. When he gets to me, he says loudly, "I'm going to show it to my mom!" Then, he holds up his quiz with his left hand. I see a "100%" written in red ink. I nod, and before I can say anything, Carlos raises his right hand, an indication that he wanted me to give him a "high five." I did and Carlos smiles.
>
> Then, Leonardo comes over, a big smile on his face, and shows us that he also got "100%." Pedro who is sitting next to me tells Carlos and Leonardo, "I got a 'good job'." All three boys smile and nod their heads. Judy, a female student, yells "Yeah!" and begins to dance, flailing her arms in the air and moving her hips from side to side. Anna, another student, looks over at Judy, who says, "I got a 100%." Groups of students begin to talk about how they did on the quiz.
>
> In the meantime, Ms. Ross is grading Hector's quiz. When she is done, she says "Holy cow," surprised (she informs me later) that Hector got a perfect score. Hector sees the quiz, smiles, raises his hands in the air, and screams loudly, "Yeah!"

The students did not display such enthusiasm in their math class. What is more, unlike in their social studies class, in math class the students were more apt to be "off task," as their teacher put it, by engaging in disruptive or distracting behavior, such as arm-wrestling, that they found more entertaining.

During the eighth grade, the students voiced frustration with their science class because they were bored of the lectures covering material the teacher said would "be on the state exam." Halfway through the semester the students had enough. They vigorously urged their teacher to ask me to help mediate the matter by compiling their concerns and suggestions—out of the teacher's earshot—and report them back to everyone involved. The teacher and I stepped into the hallway, where he informed me he was willing to have me mediate, as he hoped that would bring an end to the side conversations and note passing during class time. Wanting to aid a group of students I had come to know and respect, I agreed to mediate and met with the students while the teacher waited downstairs. Approximately 20 minutes later, I reported back to the entire class—students and teacher—without attributing any comments to any particular students. Like students in other studies (Heron, 2003; Keiler, 2011), the boys and girls at Romero indicated that they were more engaged when they were exposed to fewer lectures and more interactive, hands-on activities, such as projects and experiments. They seemed to understand that, as Barbalet (1999) explains, stimulating social exchanges can bring meaning to an otherwise boring situation. By negotiating with their teacher, the students affirmed their identities as learners and made it absolutely clear that schooling as they were experiencing it, with an emphasis on upcoming high-stakes tests, was not meaningful enough to keep them from experiencing a sense of boredom.

With both their strike in math class and their negotiation in science class, the students resisted instruction they found un-stimulating. Their resistance, however, should not be equated with the questioning of authority and academic disengagement; their defiance was not undertaken for the sole purpose of challenging their teachers. Rather, the students' resistance should be considered from Abowitz's (2000) theoretical perspective, which defines resistance as "communication; that is, a means of signaling and constructing new meanings, and of building a discourse around particular problems of exclusion or inequality" (p. 877). From Abowitz's perspective, the students at Romero were clearly communicating their displeasure with "boring" curricular activities and appreciation for more meaningful curriculum. The students' communication was much more effective when they sought mediation than when the boys went on strike. Still, the students at Romero and other students who expressed their displeasure with curriculum driven by the priorities given to high-stakes tests (see Christianakis, 2011) need to be listened to and have their concerns taken seriously by educators, educational policymakers, and politicians.

Witnessing the extent to which Romero students experienced boredom as a result of test-driven curriculum and pedagogy, I worried that students would grow to dislike schooling. What I learned was that boredom at school was influencing their college aspirations. During lunchtime, as a group of boys bemoaned the boredom they felt in their classes, one of them turned to me and asked, "Is college *this* boring, Richard?" The question caught me off guard. While I processed the query, another boy chimed in, "It's more school. What do you think? I'm not going to college." Like this boy, a number of the students equated college with more schooling and, thus, with more boredom.

Such a conclusion makes sense when one takes their schooling experience into account and considers the fact that aside from their teachers, most of them had little contact with college graduates. What makes the students' exchange so troubling is that among low-income urban minority students, "the single best predictor of 4-year college enrollment" is whether friends plan to enroll in college after high school (Sokatch, 2006, p. 128). Consequently, the high-stakes exams proponents argue will prepare students for college may result not only in situated boredom within classrooms but also curtail students' college aspirations and perhaps give rise to peer cultures that reject the pursuit of higher education.

Wanting to pique the boys' interest in attending college, I responded to the question by telling them that college provides experiences quite different from those at Romero. I pointed out that college is not as regimented and that courses are much more topical. I explained to them that there are courses in which students examine hip-hop culture, some in which the curriculum revolved around films and television shows, and still others that involved regular outings into the surrounding communities. Seemingly fascinated, they pressed me for details about these college classes and my college experiences. I abided as they interjected with commentary, such as "Wish *we* had that" and "How come *our* classes aren't like that?"

Students Deserve Better Than Boredom

The reason that some of the classes at Romero were not as interesting as the college courses seems clear—many of the students' teachers sought to live up to their obligations by narrowing the curriculum to prepare for standardized exams. In the age of high-stakes testing and teacher accountability, there is tremendous pressure on teachers not to venture beyond the district-approved curriculum and to make the most of class time by teaching test-related material. As Laguardia and Pearl (2009) explain, the "so-called reforms" resulting from No Child Left Behind (NCLB) are undermining the most central aspect of education—the teacher-student relationship, which "determines whether there will be investment and growth or whether there will be dreary days of boredom" (p. 357). In the end, much of the boredom and frustration the students at Romero experienced with the curriculum was largely due to the pressures of high-stakes tests, which they had come to "hate" and dismiss as "stupid."

While it may be tempting to dismiss the Romero students' sense of boredom as feelings associated with adolescence, such an association would be misguided. For one, students are not the only ones who experience boredom. There is ample evidence indicating that adults also experience boredom, particularly at work (Fisherl, 1993; Hill & Perkins, 1985; Mikulas & Vodanovich, 1993; Molstad, 1986; O'Hanlon, 1981; Payne, 1999), which is emblematic of modernity and (post)modern societies (Brissett & Snow, 1993; Conrad, 1997; Darden & Marks, 1999; Healy, 1984; Klapp, 1986; Spacks, 1995; Zijderveld, 1979). More pertinent still, research shows boredom can result in academically disengaged students (Dow, 2007; Fallis & Opotow, 2003; Fogelman, 1976; McGiboney & Carter, 1988; Robinson, 1975; Rubin, 2007; Wasson, 1981), lead some to cut classes (Fallis & Opotow 2003), and increases the likelihood that students will drop out (Dow, 2007; Farrell et al., 1988). These negative effects of boredom on U.S. students' academic achievement and engagement are likely being exacerbated by the practice of teaching to the test, a practice borne out of the failed accountability experiment in public education.

The unintended consequences of high-stakes testing are likely also robbing students beyond Romero of intellectually engaging activities that promote the development of higher order thinking skills. High-stakes tests cover the subject matters that research suggest may lead to the greatest sense of boredom. A study of middle school students documents boredom was "higher in classes with more abstract content (social studies, science, foreign language) as opposed to classes that deal with skill development (shop, music, gym)" (Larsen & Richards, 1991, p. 430). What is more, there is mounting evidence that exit exams increase dropout rates, particularly among underperforming students (Catterall, 1989; Jacob, 2001; Madaus & Clarke, 2001).

Politicians and educational policymakers continue to champion high-stakes tests. President Barack Obama's neoliberal educational policies, for one, are furthering the institutionalization of standardized testing as well as that of other government-imposed accountability mandates within public education (Carr & Porfilio, 2011). Hence, I fear that an observation made about Canadian public education more than 25 years ago is not only true of U.S. public education today but may also be the case going forward: "Many [students] find learning in schools an uninteresting, boring, and impersonal standardization to bureaucratic routine" (Nelson, 1985, p. 149). There are hopeful signs, however, that we can bring an end to accountability mandates and such frightful schooling. Along with the Romero students challenging curricular narrowing and the ensuing boredom, there are parents and students throughout the country refusing to take standardized tests; there are teachers subverting the testing culture by refusing to teach to the test; and there are many others also working so that schools treat students like intellectually curious, creative beings, and not like light bulbs to be sorted.

References

Abowitz, K. K. (2000). A pragmatist revisioning of resistance theory. *American Educational Research Journal, 37*(4), 877–907.

Abrams, L. M., Pedulla, J. J., & Madaus, G. F. (2003). Views from the classroom: Teachers' of statewide testing programs. *Theory Into Practice, 42*(1), 18–29.

Amrein, A. L., & Berliner, D. C. (2002, December). *An analysis of some unintended and negative consequences of high-stakes testing*. Tempe, AZ: Educational Policy Studies Laboratory, Arizona State University.

Anderson, J. E., & Ridley, T. (1978). *Cool school: An alternative secondary school experience.* Toronto: The Minister of Education, Queen's Park.

Au, W. (2007). High-stakes testing and curricular control: A qualitative metasynthesis. *Educational Researcher*, 36(5), 258–267.

Barbalet, J. M. (1999). Boredom and social meaning. *British Journal of Sociology*, *50*, 631–646.

Barksdale-Ladd, M. A., & Thomas, K. F. (2000). What's at stake in high-stake testing: Teachers and parents speak out. *Journal of Teacher Education*, *51*(5), 384–397.

Beaulieu, L. A. (1981). *Vandalism: Responses and responsibilities.* Toronto: A.P. Gordon, Queen's Printer.

Brissett, D., & Snow, R. P. (1993). Boredom: Where the future isn't. *Symbolic Interaction*, *16*, 237–256.

Carr, P. R., & Porfilio, B. J. (Eds.). (2011). *The phenomenon of Obama and the agenda for education: Can hope audaciously trump neoliberalism?* Charlotte, NC: Information Age.

Catterall, J. S. (1989). Standards and school dropouts: A national study of tests required for graduation. *American Journal of Education, 98*(1), 1–34.

Christianakis, M. (2011). Hybrid texts: Older children, rap music, and writing. *Urban Education, 46*(5), 1131–1168.

Conrad, P. (1997). It's boring: Notes on the meanings of boredom in everyday life. *Qualitative Sociology*, *20*(4), 465–475.

Corrigan, P. (1979). *Schooling the smash street kids.* London: Macmillan.

Darden, D., & Marks, A. (1999). Boredom: A socially disvalued emotion. *Sociological Spectrum, 19*(1), 13–37.

Dow, R. R. (2007). Passing time: An exploration of school engagement among Puerto Rican girls. *The Urban Review*, *39*(3), 349–372.

Fallis, R., & Opotow, S. (2003). Are students failing school or are schools failing students?: Class cutting in high school. *Journal of Social Issues*, *59*(1), 103–119.

Farrell, E., Peguero, G., Lindsey, R., & White, R. (1988). Giving voice to high school students: Pressure and boredom, ya know what I'm sayin'? *American Educational Research Journal, 25*(4), 489–502.

Fisherl, C. (1993). Boredom at work: A neglected concept. *Human Relations*, *46*(3), 395–417.

Fogelman, K. (1976). Bored eleven-year-olds. *British Journal of Social Work*, *6*, 201–211.

Healy, S. D. (1984). *Boredom, self, and culture.* Cranbury, NJ: Associated University Presses.

Heron, A. H. (2003). A study of agency: Multiple constructions of choice and decision making in an inquiry-based summer school program for struggling readers. *Journal of Adolescent and Adult Literacy*, *46*(7), 568–579.

Hill, A. B., & Perkins, R. E. (1985). Towards a model of boredom. *British Journal of Psychology, 76*, 235–240.

Jacob, B. A. (2001). Getting tough? The impact of high school graduation exams. *Educational Evaluation and Policy Analysis*, *23*(2), 99–121.

Keiler, L. S. (2011). An effective urban summer school: Students' perspectives on their success. *The Urban Review*, *43*(3), 358–378.

Klapp, O. E. (1986). *Overload and boredom: Essays on the quality of life in the information society.* Westport, CT: Greenwood.

Klein, S., Hamilton, L., McCaffrey, D., & Stetcher, B. (2000). *What do test scores in Texas tell us?* [Issue paper]. Santa Monica, CA: RAND.

Koretz, D., Mitchell, B., & Stetcher, B. (1996). *The perceived effects of the Kentucky instructional results information system* (MR-792-PCT/FF). Santa Monica, CA: RAND.

Laguardia, A., & Pearl, A. (2009). Necessary educational reform for the 21st century: The future of public schools in our democracy. *The Urban Review*, *41*(4), 352–368.

Larson, R., & Richards, M. (1991). Boredom in the middle school years: Blaming schools versus blaming students. *American Journal of Education*, *99*(4), 418–443.

Madaus, G. F., & Clarke, M. (2001). The adverse impact of high-stakes testing on minority students: Evidence from one hundred years of test data. In G. Orfield & M. L. Kornhaber (Eds.),

Raising standards or raising barriers? Inequality and high-stakes testing in public education (pp. 85–106). New York: Century Foundation.
McGiboney, G. W., & Carter, C. (1988). Boredom and proneness and adolescents' personalities. *Psychological Reports, 63,* 741–742.
Mikulas, W. L., & Vodanovich, S. J. (1993). The essence of boredom. *Psychological Record, 43,* 3–12.
Molstad, C. (1986). Choosing and coping with boring work. *Urban Life, 15,* 215–236.
Moses, M. S., & Nanna, M. J. (2007). The testing culture and the persistence of high stakes testing reforms. *Education and Culture, 23*(1), 55–72.
Nelsen, R. W. (1985). Books, boredom, and behind bars: An explanation of apathy and hostility in our schools. *Canadian Journal of Education, 10*(2), 136–160.
O'Hanlon, J. F. (1981). Boredom: Practical consequences and a theory. *Acta Psychologica, 49,* 53–82.
Payne, R. L. (1999). Stress in health professionals: Psychological and organizational causes and interventions. In J. Firth-Cozens & R. L. Payne (Eds.), *Stress at work: A conceptual framework* (pp. 3–16). Chichester, England: Wiley.
Robinson, W. P. (1975). Boredom at school. *British Journal of Educational Psychology, 45,* 141–152.
Rubin, B. C. (2007). Learner identity amid figured worlds: Constructing (in)competence at an urban high school. *The Urban Review, 39*(2), 217–249.
Shepard, L. A., & Dougherty, K. C. (1991, April). *Effects of high-stakes testing on instruction.* Chicago, IL: American Educational Research Association and the National Council on Measurement in Education.
Smith, M. L., Edelsky, C., Draper, K., Rottenberg, C., & Cherland, M. (1990). *The role of testing in elementary schools.* Los Angeles, CA: Center for Research on Evaluation, Standards, and Student Testing, University of California.
Sokatch, A. (2006). Peer influences on the college-going decisions of low socioeconomic status urban youth. *Education and Urban Society, 39*(1), 128–146.
Spacks, P. M. (1995). *Boredom: The literary history of a state of mind.* Chicago: University of Chicago Press.
Wasson, A. S. (1981). Susceptibility to boredom and deviant behavior at school. *Psychological Reports, 48,* 901–902.
Willis, P. E. (1977). *Leaning to labour: How working class kids get working class jobs.* England: Saxon House, Teakfield Limited.
Zijderveld, A. C. (1979). *On clichés: The supersedure of meaning by function in modernity.* Boston: Routledge.

|| EIGHT

Reconciling Student Outcomes and Community Self-Reliance in Modern School Reform Contexts

Brian R. Beabout and Andre M. Perry

Education: An Individual or Collective Enterprise?

Education for African Americans has historically been linked to the broad movement to improve their lot in life. Ceaselessly, from slavery and Jim Crow, toward full membership in American society, schooling was as much about academic learning as it was for ensuring the sustainability of the community in which the school was situated. Due to both *de jure* and *de facto* racial segregation of their communities and public schools, there have historically been high levels of self-determination in schooling for African Americans (Anderson, 1988). The boundaries of the racial community were often undistinguishable from the geographic communities in which African Americans lived. *Racial uplift* became the *raison d'être* in all sectors of black society, but education offered a pragmatic focus for community development, political empowerment, and economic enfranchisement. This has meant black teachers, the visible presence of the African American experience in the curriculum, and significant local decision-making power.

The latest wave of test-based reform is in direct conflict with this tradition of racial uplift and self-reliance that has substituted test score growth for community approval as the ultimate metric of educational success. Quite differently, modern test-based school reform has focused on the development of two linked areas: standards-based reform and choice (Chubb & Moe, 1990; Cuban & Usdan, 2002). From these economics-informed perspectives, students are viewed instrumentally as consumers and future workers (Strike, 2010). Schools are built to push individual success, productivity, and competitiveness. Schools improve with infusions of human capital, leading to the consequential removal of educators unable to meet assigned performance targets. African American teachers, with their concentrated presence in schools that have been deemed failing, have been the collateral damage of test-based reform. From this perspective, schools look more like corporations in which bottom lines are met, individual advancement is *merited*, and success is brought to scale. The lines between racial, school, and geographic community are demarcated to bring clarity to roles and responsibilities, but the mission of test-based reform is universal: annual score increases. This type of categorization and/or routinization reflects the social institution of the market. And while theories of efficiency and scientific management entered the education policy vocabulary long ago (Thorndike, 1904; Tyler, 1950), never has there been such political acceptance or structural alignment for the use of student testing as the sole metric for educational decision making at the school, district, state, and federal levels.

This current pervasiveness of market approaches is reflected in the reform language of state takeover, school turnaround, and reconstitution. As a consequence, since 2001, administrative control of many schools serving students of color has shifted from local educators and elected school boards to the states and the federal government, who set the accountability policies and determine student and school accountability rules based on test scores. The following chapter interrogates this facially benign policy of raising student achievement with respect to the potential impact on the legacy self-determination of African American schooling.

Divergent Perspectives

These changes have fueled a heated public debate about schooling for the poor in America. This debate includes neo-liberal school reformers who tend to favor school choice, charter schools, and a heavy emphasis on state testing for school evaluation (Beabout & Jakiel, 2011). Opposite this camp are the community control advocates that tend to favor increased public investment in schools, community-driven decision-making, and multiple measures of school evaluation. Thomas (2011) has offered a similar dichotomy of school reform camps, dubbing them the *no excuses* reformers and the *social context* reformers. And while these binary divisions are necessarily simplistic and fail to capture the nuances of the views of individuals and groups who care about improving schools, these divisions foreground an important question in our current reform debates: *What are the consequences of our current obsession with creating schools that can raise test scores?* What this debate currently lacks is a recognition of the competing values that drive each camp. The neo-liberals emphasize student academic achievement over community self-determination while the community control camp does not. Both views have inherent strengths and flaws.

This chapter provides a theoretical examination of the impact of these differing sets of values (student outcomes vs. community self-determination) and suggests a conceptual road map for improvement based heavily on our work with public schooling in post-Katrina New Orleans, perhaps the American city where test-based accountability has been implemented in its most philosophically pure form (Beabout, 2010; Perry & Schwam-Baird, 2010).

The Rise of the Testing Regime in New Orleans

In post-Katrina New Orleans, the student outcomes reformers have reigned almost unopposed. While public education in New Orleans is a 160-year story filled with remarkable feats of self-determination and collective action (DeVore & Logsdon, 1991), a more recent history of the district reveals decades of poor academic performance, limited public investment, and systemic mismanagement (Roesgen, 2005). In a situation ripe for a revolution, the neo-liberal philosophy of prioritizing student achievement on state tests over the other functions of schooling (see Rothstein, Jacobsen, & Wilder, 2008) has been supported by a pro-market state government and bolstered by infusions of cash from market-oriented foundations and the federal government (Ritea, 2005; Simon, 2007). Since the storms of 2005, scores on state tests have risen (Perry & Schwam-Baird, 2010) while complaints about increasing racial segregation (Abramson, 2010), illegal firing of teachers (Robertson, 2012), and the exclusion of special needs students (Ritea, 2007) have generally been marginalized to small pockets of activists with little access to decision-making channels. Entranced by the widely held American belief that the product of good schools is good test scores, the media has showered much attention on the reforms in New Orleans. And while there have been many notable successes, the emerging gulf between our public schools and the predominantly African American community that they serve gives us pause.

It is undeniable that test-based accountability has limited the self-determination of the local African American population in running their own schools. The publicly elected school board has lost control of many schools and a state-appointed superintendent holds the most influential leadership role in the city. Meanwhile, rapid expansions of New Leaders for New Schools, Teach For America, and their affiliated charter management organizations have flooded the employment market with young teachers and school leaders from out of state. While many local educators still work in the schools, there has absolutely been a shift of power away from New Orleans' African American community and the once-mighty teachers union that for years influenced politics in the city. This is further evidenced in the recent election of New York native and Teach For America's New Orleans chapter executive director Kira Orange-Jones to the state education board with significant campaign support from state and national-level luminaries (including New York City mayor Michael Bloomberg) (Vanacore, 2011). This local development has led us to consider the question: *Are there times when limiting peo-*

ple's self-determination can be ethically justified? This analysis of the relationship between test-based accountability and urban communities begins with a look at the history of self-determination in African American thought and education. While educational self-determination is today being challenged in many low-income communities with many different racial compositions, we focus on the African American experience here because that is the context of such change in our city.

Historical Roots of Self-Determination

In the Americas, the period of Reconstruction saw a particular social and political activism conducted by and on behalf of the formerly enslaved people of African descent. The term *racial uplift* and its derivatives (*upliftment, community control*) are synonymous with the quest for civil and political rights by the descendants of the enslaved. Forged out of the crucible of slavery, racial uplift as a movement shaped black activism and the black education agenda. The backdrop of slavery accentuates why proponents of uplift do not separate freedoms such as education from political, economic, intellectual, and religious rights. As the Brookings Institution has reported:

> The moral, legal, and rhetorical pursuit of collective rights of access was but an essential strategy in a multifront war for a much larger prize: the uplift of a people, once mostly enslaved, afterward still widely despised. . . . Indeed, black leaders and organizations have always known that they must pursue the vast and varied interests of their stigmatized and marginalized constituents by any realistic mechanism available. Rights were more a means than an end. (Foreman, 1998)

Du Bois' oft-referenced citation underscored the overarching problem to be solved through uplift: "The problem of the twentieth century is the problem of the color-line" (Du Bois, 1989, p. 10). The African American community had little choice but to respond *as a community* to deal with the legacy of racial discrimination, legal segregation, and slavery. Consequently, racial uplift has left an indelible imprint on black education as an ideology and field of study (King, 2005).

Notwithstanding the civil rights work throughout the world prior to Reconstruction, but certainly since, activists have pursued quality education and schools as an essential (but not singular) vehicle for racial and gender equality, full citizenship, and restorative justice. This work includes a specific focus on civil and educational rights, curriculum, instruction, and claims for physical school buildings. Even prior to the legal end of slavery, black educationists fought for state-sponsored universal education when white landowners felt the state had no right to impose education. Plantation owners and industrialists also did not want to disrupt the social arrangement developed after more than 200 years of slavery (Anderson, 1988). Consequently, African American public education is a byproduct of racial uplift.

Again, schools, diplomas, and test scores were not the end goals. Early proponents focused on loftier goals such as community/self-sufficiency. Self-reliance and community determinism are manifested in activists' pursuit for educational space, both physical and intellectual. Angel David Nieves's interdisciplinary work on "African Americans' struggle to claim a 'space' for themselves within the complex social, economic, and political fabric of U.S. dominant culture from the antebellum era through the present" is a good framework to view self-determinism in black education (Nieves & Alexander, 2008, p. 1). This work shows that black education was also about taking stock in the physical aspects of communities, neighborhoods, or cities. However, during Reconstruction and Jim Crow, blacks had to claim land, businesses, and schools without the start-up capital to do so. Partnership, government sponsorship, and philanthropy would provide the seeds toward universal education and self-empowerment (Anderson, 1988). Interestingly, black activists have always had to negotiate the principle of uplift while working with external foundations, religious organizations, and individuals who sponsored the uplift agenda. Thus, we can see black educational self-determination not as a simple description of who is involved in education of black children, but more accurately, who is making the decisions regarding the enterprise.

Similarly, Mary S. Hoffschwelle's (2008) examination of the Rosenwald school building program epitomizes idealized attempts of philanthropists and blacks working together to increase the capacity of racial communities. Booker T. Washington teamed with Julius Rosenwald, president of Sears, Roebuck and Company, to establish the initiative. "The most important element in that coalition was African American activism and self-help, mandatory components of the building program that Black southerners seized upon to make these schools their own" (Hoffschwelle, 2008, p. 278). The facility campaign met modern standards of building construction. All the sites had to be deeded to the local board of education or the state. Black school patrons contributed materially to the design and construction and believed in "the social and personal value of universal education and wanted white southerners to accept that black education was a public responsibility" (Hoffschwelle, 2008, p. 282).

In addition to buildings, self-determination could not be achieved without a superior education in terms of curriculum and instruction. W. E. B. Du Bois certainly did not seek education merely to achieve economic equality. As a supporter of the American Missionary Association's mission to build black schools in the liberal tradition, Du Bois wanted them to focus on the training of black teachers and higher education. As he articulately describes in his own words:

> The teachers in these institutions came not to keep the Negroes in their place, but to raise them out of the defilement of the places where slavery had wallowed them. The colleges they founded were social settlements; homes where the best

> of the sons of the freedmen came in close and sympathetic touch with the best traditions of New England. The lived and ate together, studied and worked, hoped and harkened in the dawning light. In actual formal content their curriculum was doubtless old-fashioned, but in educational power it was supreme, for it was the contact of live souls. (Du Bois, 1989, p. 70)

The sentiment that black education was "not to keep the Negroes in their place" is the recognition that education without racial uplift falls short to the larger need of Americans disenfranchised by slavery. Harvard educated, Du Bois unashamedly promoted a classically liberal, northern education. However, he was unsatisfied with the acquisition of skills to *fit* into society or to serve a role. This was made clear by his rejection of industrial education, which sought to conveniently transition blacks from a slave and agricultural economy built on slavery to an agricultural one built on free men. "'What they [formerly enslaved] desire is assistance without control.' The values of self-help and self-determination underlay the ex-slaves' educational movement" (Anderson, 1988, p. 5).

The goal of racial uplift continued throughout the 20th century. Again, racial uplift was not compartmentalized neatly into education, economic development, health care, etc. The cross-pollination of these factors converged, especially in schools. For instance, organized labor has deep connections in education. In Milwaukee, local Urban League administrator William Kelley used schools to protect the rights of community members, for "the public schools not only were potential employers of middle-class black teachers but also served as key institutions that socialized black youth into their roles in the city's racialized labor market" (Dougherty, 2004, p. 13). Disenfranchised groups naturally partner across sectors to mobilize and share scarce resources and to form political alliances that empower the collective whole (Stone, 2002). However, sector crossing creates opportunities to prioritize values, which exposes internal and external competing interests.

The need for community determinism and community control reaches impasses when blacks are asked to place their professionalism, gender, or region over their racial identity. Teacher unionism during the 1968 NYC teacher strikes revealed tensions as black community mobilization presented a threat to professional inclusion of black educators. Twenty years prior, the United Federation of Teachers led a desegregation effort and voted not to charter segregated locals. During the high-stakes strike, self-help and community control sounded like aggressive Black Panther rhetoric to the union (Perlstein, 2004). In the moment of the 1968 strike, the union saw community control as a backwards step moving away from the ideal of integration. For many in the African American community, integration meant losing ground on school leadership, curriculum controls, and access to power.

Collective bargaining has been an essential tool for racial uplift particularly for blacks in service, manufacturing, and professional industries. However, the battle to protect or erode collective bargaining has ensued since the passage of the National Labor Relations Act in 1935. Ever since, racial politics have been entangled in labor disputes. Attempts to decertify unions in Wisconsin and in New Orleans, Louisiana, provide examples that will be highlighted below. However, this intertwined network of political forces supporting racial uplift is almost unrecognizable in the recent rise of achievement-centered management and their brand of market philosophy being used to improve schools. The history described above provides useful context for understanding the present-day resistance to test-based reforms that is common, though by no means universal, in African American communities. The following discussion outlines and critiques the basic positions of two primary constituencies in the current reforms debate. We will call them the *community control camp* and the *student outcomes camp*. The former traces its roots quite directly to the racial uplift narrative, the Rosenwald Schools, and Du bois' brand of explicitly raising future generations to change society, rather the simply fit in. The latter finds its roots in scientific management and the neoliberal philosophy that the social ills of poverty, inequality, and racism can be addressed through market-oriented reforms based on maximizing test-score achievement for low-income students of color.

Historical Roots of the Student Outcomes Camp

Beginning with the massive growth in public school attendance following World War II, but perhaps best encapsulated in the 1983 *A Nation at Risk* report, the project of providing public education in the United States has become less about providing access and more about producing measurable results for public investment. Investments at the federal level initially ballooned with the Elementary and Secondary Education Act of 1965 that has become the No Child Left Behind Act of 2001 (and is currently undergoing revision by Congress). While federal funding still makes up only about 10% of education funding nationwide, the leverage of this money has prompted states to establish content standards, create annual achievement tests, and devise reform plans that have tended to move decision-making away from African American communities and toward the states and their nonprofit partners.

The Modern Student Outcomes Camp

Since the entry of the federal government as a major player on the U.S. public education scene in 1958, the struggle for improved schooling has expanded on ideological and philosophical grounds. Particularly in urban settings, black and brown children are the focus of the student outcomes camp that emphasizes choice, market approaches, charter schools, vouchers, and portfolio management. Libertarians, neoconservatives, and neoliberals have found common ground in this camp under the umbrella of the broader choice movement (DeBray-Pelot, Lubienski, & Scott, 2007).

"Choice" has been an undeniably unifying battle cry. "And although many see choice as a quintessential value in consumer-oriented American society, some conservatives and progressives champion choice as a means of empowerment for disadvantaged communities disenfranchised by that society—the 'new civil right'" (DeBray-Pelot et al., 2007, p. 205). But to say everyone is for choice has not quelled the contentious battles around charter schools, vouchers, state takeover, and other initiatives. In particular, market-driven, test-based strategies have led to labor disputes that make it difficult to differentiate if management seeks to eliminate collective bargaining or to improve educational outcomes. These actions clearly challenge the uplift agenda. Nevertheless, a discussion of key market principles in education under the umbrella of "choice" is warranted.

Testing and Choice Proponents

Standard-based curriculums, high-stakes testing, deregulation, decertification, charter schools, and vouchers have expanded mightily over the last 20 years under the hydra of autonomy, choice, and competition. "Such policies often blend public and private provision, with public components increasingly staffed by conditional employees rather than career civil servants or union workers, subject to contractual accountability for outcomes, heavily focused on numerical data for bottom-line accountability, and subject to sharp fiscal management discipline from above" (Bulkley, Henig, & Levin, 2010; Menefee-Libey, 2010, p. 57). These are the hallmarks of the student outcomes camp.

In response to *A Nation at Risk* in 1983, market theorists John Chubb and Terry Moe (1990) posited that the incremental reform movement of the day was incapable of reforming itself because of thick bureaucratic processes, stifling within-school regulation and immovable governing boards. They sought freedom from the internal and external bureaucracies to allow for innovation, leadership, and rapid change. They found difficulty untangling educational outcomes from the confused organizational and governmental context in which schools resided. This opened the political window for the modern accountability, deregulation, decertification, and conditional employee movements. Chubb and Moe did not specifically recommend these issues, but like-minded market-driven thinkers now had entre into the education reform world. Significant also is a response to Chubb and Moe by Wells (1991) in which she identifies, presciently, inequitable access to desirable schools, increased racial and socioeconomic segregation, and diminished democratic governance as negative consequences of their purely hypothetical proposal for relatively unrestrained school choice.

The passage of No Child Left Behind (NCLB) in 2001 signifies for many the start of the test-based accountability era. Republicans' proclivities for government accountability and performance ushered high-stakes testing, state takeover, and expanded opportunities for charter schools. Educational researchers had extensive evidence on the size and nature of the achievement gap between different ethnic groups. However, the federal government charged states to close it. A federal reach into education is usually discouraged by the Republicans (Marshall, 2011). However, it did embrace an affable business orientation to reform. Particularly, gap-closing involved psychometric data collection and decision-making, students' required acquisition of critical subject areas, deregulation, autonomy, school turnaround, as well as dramatic changes in workforce and talent. What is not emphasized in market approaches are community development, collective bargaining, racial uplift, and strong public oversight (elected governance).

Over the last 20 years, CEOs of some of the most successful businesses involved themselves in public education debates. The Bill and Melinda Gates Foundation, the Broad Foundation, Fisher Foundation, and Walton Family Foundation have collectively contributed hundreds of millions of dollars to organizations that embraced market principles or who were at least aligned with philosophies that made their businesses successful. Their foundations have contributed millions to education reform (Reckhow, 2010).

For instance the Walton Family Foundation's "core strategy is to infuse competitive pressure into America's K–12 education system by increasing the quantity and quality of school choices available to parents, especially in low-income communities. When all families are empowered to choose from among several quality school options, all schools will be fully motivated to provide the best possible education. Better school performance leads, in turn, to higher student achievement, lower dropout rates and greater numbers of students entering and completing college" (Walton Family Foundation, 2011). In 2010, they invested $157 million in education reform including the *Investment Strategies* of public charter school choice, private school choice, district reforms, particularly open enrollment and district school choice, and cross-sector parental choice, where parents are empowered to choose across school sectors.

Similarly, the Broad Foundation's focus on market results is reflected in their "Approach to Investing: Venture Philanthropy. . . . We take an untraditional approach to giving. We don't simply write checks to charities. Instead we practice 'venture philanthropy.' And we expect a return on our investment. . . . We hold ourselves and our grantees accountable for results—because results are what will improve the education of every American student" (Broad Foundation, 2011). The return on investment is measured using test scores and other forms of discrete, easily collected information. Community wellness is not measured as a re-

Table 8.1. Community Control Camp

Axioms of the Community Control Camp • Urban education is about social uplift for poor communities of color. • Social uplift is primarily the work of minority communities, with supporting roles for other members. • Social uplift can be effectively addressed only through coordinating education, health care, public safety, housing, etc. • Urban educational problems require both school and non-school solutions. • Making strong communities is the goal of education and the source of metrics for success. • Teachers, parents, and community groups must work together to "bring up" our children (seeking optimal outcomes for all learners).
Strengths of the Community Control Camp • Community resources are drawn into the work of schools. • Broad definition of success, more holistic, inclusive. • Educational success tied to community success. • Community self-determination seeks to support local culture and prepare students for the global economy ("I know who I am and what I can do").
Weaknesses of the Community Control Camp • Self-determination means "local school boards" to many policymakers and reformers, and this is often connoted to mean "status quo." Alternatives need to be provided (Warren, 2005). • Simply defining and sustaining *self-determination* and *community* in diverse urban contexts is challenging. • Self-determination where there is a lack of local capacity leads to failure.

sult. As stated earlier, market approaches are not fond of bureaucratic structures that increase inefficiencies. This aligns with conservative values.

In an ironic twist, the conservative Heritage Foundation eschews state intervention, for "the Constitution does not provide for a federal role in education, and public schools have traditionally been under the jurisdiction of local authorities. Washington's intervention seems to have brought out the worst in education governance. It has led to ever-increasing spending and bureaucratic bloat while undermining schools' direct accountability to parents and taxpayers" (Marshall, 2011, p. 1). This statement should not be taken as support for self-determination, however, but rather as a vote for the individual's right to choose in the marketplace educational vouchers.

While the overall investment of these groups is relatively small compared to public education spending, the impact of millennial choice advocates has been felt. Terms and phrases familiar to corporate America such as *human capital*, *ven-*

Table 8.2. Student Outcomes Camp

Axioms of the Student Outcomes Camp • Urban education is about closing the racial achievement gap. • Social inequality can be effectively addressed through closing the achievement gap. • Urban educational problems can be solved at the school site. • Achievement of state standards and college graduation are the metrics of success. • Teachers and schools must get all students to meet state standards at all costs.
Strengths of the Student Outcomes Camp • Teachers and schools that do not generate academic achievement gains will not operate very long. • School choice is advocated—students are not stuck in failing schools. • Accountability mechanisms focus on student achievement.
Weaknesses of the Student Outcomes Camp • Schools are often forced to prioritize math and English over other important educational goals. • Local community tends to be seen through "deficit lenses" and this strains community-parent relationships. • Teachers become service-providers, not adult mentors, which may contribute to less-satisfying work for teachers and increased turnover. • Reform does not take up the broader inequalities of society: *"no curricular project can fundamentally transform knowledge and its distribution if it is not part of a process of transforming social relations as well"* (Perlstein, 2002, p. 249).

ture capital, *start-up*, *scale-up*, *turnaround*, *incubation*, *management organization*, and *data driven decision-making* emerged quickly in the education landscape. School leaders have also adopted nomenclatures such as *CEO*, *COO*, and *CFO* in lieu of *superintendent*, *principal*, and *treasurer*. As titles and phraseology has changed, so too has school culture. Common to the drivers of the new school market reform is the "laser-like" focus on closing the achievement gap. In and among themselves, the aforementioned ends of market driven approaches are not necessarily an affront with black individuals. Who doesn't want students to perform better? Rather, it's the black community's demand for a racial uplift focus that is in conflict.

A Comparison of the Two Camps

Both the community control camp and the student outcomes camp can be described by a set of axioms that, while no means exhaustive, firmly show their philosophical position and make clear how they reach their conclusions relating to the control and purposes of schools. In practice, both views have certain strengths and weaknesses. These are outlined in Tables 8.1 and 8.2, but discussion is forgone given space limitations here.

Black Educators vs. Education Reformers

In the current reform debates, the rhetorical convening power of choice hides nuanced, clear divisions between both camps who are ostensibly working on behalf of the disenfranchised. It is important to clarify the areas of contrast that exist in a somewhat blurry convergence of advocates supporting the amorphous concept of choice.

The inherent objectives of the black educationists' goal of self-reliance as a by-product of place-making through schools comes in direct conflict with choice advocates' fight for deregulation, decentralization, and decertification. Universal, public education has been inextricably linked to labor, housing, and busing movements. In addition, the outcomes camp (with their emphasis on closing racial achievement gaps) seeks to rid school of links to racial uplift agenda (organized labor, universal schooling, strong electoral leadership).

Historically, black recipients of the various educational movements wanted the freedom to create their own pedagogies, rules, and schools. We should remember the caution that W. E. B. Du Bois noted about the Freedmen's Bureau. He believed that the bureau certainly set forth a system of free labor, free public schools, and black businesses. However, "it failed . . . to guard its work wholly from paternalistic methods that discouraged self-reliance" (Du Bois, 1901). It is this insistence on self-reliance, even when the spoils of paternalism are deeply desired, that creates the enduring rift between the two camps described here. It allows, and in fact necessitates, black educators from the community control camp to respond negatively to even the best of the victories of the student outcomes camp. In New Orleans, one can sit in a room where positive test score growth is displayed by student outcomes reformers and witness those in the community control camp focus exclusively on poking holes in the numbers by referencing the number of expulsions, the limited special education services provided, the challenges of city-wide school enrollment, and decentralized busing challenges. Improvement, even improvement that is desired by both camps, will not be accepted by community control advocates when the work is *owned* by student outcomes advocates. There is a lack of trust and a lack of collaborative, across-the-aisle work that has become the hallmark of much of recent U.S. politics. Both groups will have to make concessions in their proposals and in their style of interaction if this is to change.

Loss of Community Control: The New Orleans Context

As the outcomes camp ascends to power, conflicts around organized labor continue to be a source of racial conflict. The aftermath of Katrina started a new chapter in the saga of those who embrace racial uplift amid competing values and agendas of movements that explicitly share the same goal. Specifically, black educators and others who are proponents of racial uplift struggle to find connections with the student outcomes camp whose rapid (and contested) ascendancy to power in New Orleans damaged the potential for collaborative work between the two groups.

Labor conflicts in New Orleans reached an apex in the month following Hurricane Katrina. Without residents and their tax revenues flowing into the district, an already cash-strapped New Orleans School Board cancelled paychecks and insurances for its employees. The employees, approximately 70% of whom were black, including teachers and custodial and cafeteria workers, bus drivers and support staff, received one paycheck on September 1, 2005. Without concrete information on many of the employees' whereabouts, hundreds of workers did not receive written notification of the cancellation of pay. Many employees received word about the decision from news media outlets, union groups, and word of mouth. The financial and collegial blows of the pay cancellations added insult to injury to an already displaced group of working- and middle-class employees. The teachers who did return and attend subsequent open meetings of the Orleans Parish School Board (OPSB) directed their anger toward its members instead of on the reform. However, not receiving pay was a critical decision that preceded another political move, which set the table for new educational delivery system.

During an emergency session of the Louisiana legislature after the storm, the passage of Act 35 expanded the state's authority to take over schools by redefining what was considered failing. The state-run Recovery School District (RSD) assumed control of the failing schools. In addition, the RSD served as a school authorizer, giving it ability to create new charter schools. Shortly thereafter, the federal government issued Louisiana approximately $45 million over the next two years for the development of charter schools. As school authorizers, the Louisiana State Department of Education through its charter school office and the RSD could award approved educational service providers with start-up grants and services. Act 35 initiated the power shift from the New Orleans School Board to a new swath of nonprofit school providers. It also created new market conditions for service providers that are external to the schoolhouse. The state's charter office also increased its profile and influence on the how schooling would be delivered. Dissent along the lines of self-determination and community control naturally followed.

Two of four black members of the OPSB opposed the changes. School board president Torin Sanders and superintendent Ora Watson were on the losing side of a 4–2 vote to create a charter district in Algiers. The board president and superintendent sought to retain 4 of the 13 schools in Algiers that would be eventually placed under the auspices of the RSD. Sensing a loss of local control, a group of mostly black business leaders, activists, and religious leaders successfully filed an injunction to stop a plan to transfer control of the schools to the charter managing organization. The injunction eventually expired. In addition, Alvarez and Marsal, the accounting firm that managed and supervised the board's finances, informed members that they did not have the fiscal resources to open the schools. The opposition to the portfolio model was staged as the fight for local control. The charter district would eventually become the Algiers Charter School Association (Bogotch, Miron, & Biesta, 2009).

The Louisiana Board of Elementary and Secondary Education's (BESE) takeover of 112 of 127 New Orleans schools and the subsequent placement of those schools in the RSD led to the OPSB officially terminating the contracts of all of its employees. Employees went from not receiving pay to not having a job with the NOPS district. A bankrupt school board without control of 88% of its schools simply could not continue to employ its teachers, custodial and cafeteria workers, bus drivers, as well as other support staff. The new nonprofit providers possessed the autonomy to hire and fire their employees, with no obligations to provide preferential treatment to previous school system employees.

Consequently, United Teachers of New Orleans (UTNO), the local teachers union, fought vigorously against the reforms that led to the termination of teachers. UTNO started a statewide campaign titled "Refuse to Lose." President Brenda Mitchell stated, "We've been painted as obstructionists, which is not the case . . . I've been told that the union is trying to run the schools. All we are saying is that we should work together on this" (Tisserand, 2007). On June 30, 2006, the UTNO contract with NOPS expired, and the school board did not vote to renew the contract. With a significant reduction in membership due to the transfer of schools to the RSD and without a collective bargaining agreement, the union saw its influence significantly reduced.

From the perspective of the community control camp, the loss of collective bargaining was viewed as a deadly blow to black empowerment, racial uplift, and/or community control. The outcomes camp chooses explicitly to seek improvements in public schools, but without recognition of the broader goal of racial uplift. The student outcomes camp's goal of gap closing unapologetically eschews the racial uplift agenda items such as organized labor and economic development. In essence, the outcomes camp places emphasis on improving the academic performance of poor children of color while removing any agenda for reducing other areas of inequality that face both children and adults. Strategically, this constricted set of concerns is politically savvy as it has garnered support from frustrated educators as well as the base of the Republican Party, which has steadfastly opposed government-led attempts to address the myriad inequalities that exist in American society, relying instead on the private sector to do this work, when it is profitable. This narrowed strategy also allows the student outcomes camp to work with a smaller set of goals, making it somewhat easier to demonstrate success to funders and voters. For the traditional African American civil rights organizations, narrowing the focus of their work to student outcomes only is a contradiction. They fight for expanded rights and great equality for all members of their respective groups. One African American group, however, has managed to successfully launch a platform that narrows the struggle for equality to educational outcomes. This is a move that has created much animosity but also an interesting study of political maneuvering within the tensions between the students outcomes and self-determination camps.

A Contentious Crossroads: The Black Alliance for Educational Options

The Black Alliance for Educational Options (BAEO) offers a unique but inevitable progression in the student outcomes vs. self-determination debate. Whereas many African American civil rights organizations, educational organizations, and community groups have sought to claim social, educational, and political space in the public schools, BAEO espouses broader self-deterministic principles that transcend public schools but are limited to education. BAEO seeks "to increase access to high-quality educational options for Black children by actively supporting parental choice policies and programs that empower low-income and working class families" (Black Alliance for Educational Options, 2012). BAEO's promotion of high-quality school options has led them to advocate for nonpublic options as well as charter schools. Their advocacy for racial communities separate and apart from traditional educational communities historically promoted by black mainstream groups places BAEO at a contentious crossroads.

The Louisiana chapter of BAEO has been on the frontline of reform on behalf of the state. The national president of BAEO, Kenneth Campbell, was the former charter school director for the Louisiana Department of Education. Louisiana BAEO was the public face of Republican Governor Bobby Jindal's education legislation that led to the State Department of Education's Student Scholarship for Excellence program, which is the newly created voucher program for the state (Barrow, 2012). Vouchers may be the most obvious derivative of the market approach to schooling and the most challenging reform initiative to the black educator tradition. However, BAEO's cheerleading for black school leadership as well as economic and educational freedom creates a libertarian perspective in the black political continuum. If the Louisiana style of school reform is to succeed nationally, groups like BAEO that can peel off traditionally Democratic-voting African Americans will play a key role. What is uncertain is whether their unique political positioning will cost them support in African American communities where their objectives include expanding private school vouchers and using student testing to diminish local control of public schools. On one hand, BAEO's emphasis on moving black students into the private system (which could ostensibly be black-controlled) might be the ultimate move toward community control. On the other, a weakened public system in districts where the vast majority of families are people of color increases the likelihood of state takeover and loss of community control in the public system. How their work is perceived is unknown as they are relatively recent entrants into the mainstream educational reform dialogue. For more road-tested thoughts on how marginalized racial groups might manage the education of their own young people, we next briefly look at educational thought from the Native American perspective.

Transsystemic Schools: Lessons from Indigenous Education

It should, at this point, be clear that testing advocates have used the current wave of educational reform to limit local control of certain schools on the argument that traditional district governance has failed to adequately educate students. Self-determination has been a historically assumed reason for education. There is a long-standing cultural ethic of self-determinism in black communities rooted in efforts to free the enslaved and in the quest for equality after the official end of slavery. Modern reforms, however, buttressed by widespread acceptance of the idea that the job of schools is to produce high-testing students, have begun to erode this self-determination. And the communities most directly affected are poor communities of color, for whom the right to self-determination has been a multigenerational, and ongoing, struggle. How, then, are we to proceed when there is a trade-off between educational performance and local community control?

In the African American tradition, community empowerment has been indistinguishable from individual empowerment. The gains and pitfalls in education reflect the struggle for control of public and private assets. A critical (though generally unasked) question of this debate is: *How are the social costs on community self-determination accounted for in the current wave of market-based reform?* In a liberal democracy, when do individual rights (or the responsibilities of the state) trump communities' rights to self-determine? What is the purpose of reform if communities are not able to self-determine? *Are the state's actions to address the lack of educational achievement for poor children of color merely a convenient way to maintain an unequal social order by limiting communities' struggle to self-determine?* Given historical patterns of enslavement and acute racialized oppression in our country, these questions cannot be taken lightly. In what follows, we extrapolate some lessons learned from the educational programs designed for Native Americans over the last 150 years.

After the 1880s, American cultural genocide of indigenous people was attempted primarily by the use of government-run boarding schools that sought to "de-Indianize" Native American children, thus breaking the cultural links that might facilitate future resistance to Anglo domination on the continent. Between 1880 and 1900, the government increased its spending on government schools 40 times over, and had created a network of 147 reservation day schools and 106 boarding schools (Adams, 2008). Local control in the education (and supervision) of children gave way to state control with the explicit intent of cultural genocide. In the documentary film *We Shall Remain,* former boarding school student Walter Little Moon (Oglala Lokota) notes:

> This is the education that was promised us, that was *guaranteed us,* through the treaties. But it wasn't. It was torture. Brainwashing. They called us many different names: savage, dumb. We got beat for *lookin'* like an Indian, *smellin'* like an Indian, even *speakin'* Indian. Everything I did . . . (wipes away tears). (Grimberg, 2009)

These federal indigenous education policies claimed legitimacy by offering to assimilate *savage* children into mainstream American society through basic education, re-culturation, and the teaching of English and Christian spiritual beliefs. In the long run this program thankfully failed, but the costs were horrible on students, families, and native communities that lost bonds of kinship that are impossible to repair.

On the surface, current test-based educational reform does not explicitly seek to eliminate African American culture in an effort to bring the students into the mainstream. But in practice, is the current emphasis on removing educational decision-making from poor communities of color via testing and choice reforms any different? Has educational testing (rather than rifles) been used as a tool to sustain (or even grow) the socioeconomic gulfs that exist between communities in the United States? Buras (2010) identifies the loss of teaching jobs as a severe economic blow to the black middle class in New Orleans. A June 2012 state court decision found that the firing of nearly 7,500 Orleans Parish School Board employees in the wake of the storm was implemented illegally, paving the way for damages to be paid to many of the fired teachers (Robertson, 2012).

What lessons can be learned for the education of low-income communities of color in the United States? Borrowing from the literature on indigenous knowledge and education (Battiste, 2008; Madjidi & Retoule, 2008), and the U.N. Declaration of the Rights of Indigenous Peoples (2007), the animating spirit guiding the reform of our systems of public education is a balance of power between local communities and national (undoubtedly Eurocentric) priorities. If the United Nations was compelled to pronounce that indigenous people have the right to establish and administer education for its children, why is it that poor communities of color in the United States are seeing their rights diminished? The true answer to this question lies at the intersection of our American belief in tests as the ultimate measure of educational performance (Tye, 2000) and our physically and symbolically violent insistence on assimilation as the ultimate act of building a nation. To overcome these harmful tendencies, we need to create new understandings of knowledge and education that are "'transsystemic' . . . and reach beyond two distinct systems of knowledge [indigenous and Eurocentric] to create fair and just educational systems and experience" (Battiste, 2008, p. 90). As test-based accountability policies use a Eurocentric view of the world to strip educational self-determination from non-dominant communities, a *transsystemic* view of accountability would combine state and locally defined educational outcomes, track progress toward ensuring that students can both succeed in the mainstream economy and sustain the cultural linkages of their community (Delpit, 1995) view educational and other social goods in relationship to each other, and ensure that the local community has authority in educational decisions.

In practical terms, this means ensuring that local citizens are encouraged to work in education; this means that when schools are opened or closed, the local community is a part of the decision; this means that school leaders must be chosen by parents and community members so that they can effectively integrate the education of children with other community needs (Khalifa, 2012), and that student testing is only one piece of a larger collection of data used to make formal decisions about school performance, which might include community involvement, site visitation, alumni performance, and student social and emotional growth (Rothstein, Jacobsen, & Wilder, 2008; Schwartz, Hamilton, Stecher, & Steele, 2011).

Test-based reform policy has led to a loss of community control of public schooling in many poor communities of color in the United States. The historic legacy of strong community control for African American communities has reached a crossroads. There is bipartisan political support for the idea that poor test performance justifies the loss of local control via reconstitution, state takeover, and the expansion of quasi-private charter schools and, less strongly supported, private school vouchers. This portends more cities where such loss of self-determination will occur. Even if test scores rise under such a system, the idea that a community is deemed unable to educate its own children should shock the conscience. Lessons from the history of Native American education (as well as indigenous education worldwide) teach us that the unilateral imposition of Eurocentric schooling on marginalized communities, regardless of outcomes, is an act of cultural genocide. While the varying components of a successful educational program for poor communities of color in the United States are persuasively described elsewhere (Gay, 2000; Ladson-Billings, 1994), the essential lesson of the discussion is that communities have an inviolable right to manage the public institutions that directly affect them. When testing policies are used to wrest control of public schools from local communities and place that authority in the hands of educational technocrats or distant policymakers, a disservice has been done. Policymakers faced with this understandably difficult decision are urged to reject the simplistic urge to do whatever it takes to achieve annual test-score gains, but to consider the impacts to our democracy when our poorest and more marginalized citizens have access to less democracy than their wealthier and whiter neighbors. The reform pendulum (Cuban, 1990) feels to be reaching an apex quite far from the transsystemic view offered by Battiste (2008). Those interested in putting such ideas into practice are wise to organize now, so that they are prepared when the window opens.

References

Abramson, L. (2010). *Parents push for diversity in New Orleans' schools.* Retrieved August 7, 2012, from http://www.npr.org/templates/story/story.php?storyId=129531693

Adams, D. W. (2008). Fundamental considerations: The deep meaning of Native American schooling, 1880–1900. In M. Villegas, S. R. Neugebauer, & K. R. Venegas (Eds.), *Indigenous knowledge and education: Sites of struggle, strength, and survivance* (pp. 85–91). Cambridge, MA: Harvard Educational Review.

Anderson, J. D. (1988). *The education of blacks in the South, 1860–1935.* Chapel Hill: University of North Carolina Press.

Barrow, B. (2012, April 5). *Legislature gives final approval to package of bills to remake education in Louisiana.* Retrieved August 8, 2012, from http://www.nola.com/politics/index.ssf/2012/04/legislature_gives_final_approv.html

Battiste, M. (2008). The struggle and renaissance of indigenous knowledge in Eurocentric education. In M. Villegas, S. R. Neugebauer, & K. R. Venegas (Eds.), *Indigenous knowledge and education: Sites of struggle, strength, and survivance* (pp. 85–91). Cambridge, MA: Harvard Educational Review.

Beabout, B. R. (2010). Leadership for change in the educational wild-west of post-Katrina New Orleans. *The Journal of Educational Change, 11*(4), 403–424.

Beabout, B. R., & Jakiel, L. B. (2011). Family engagement in charter schools. In S. Redding, M. Murphy, & P. Sheley (Eds.), *Handbook on family and community engagement* (pp. 147–151). Charlotte, NC: Information Age.

Black Alliance for Educational Options. (2012). Our mission and beliefs. Retrieved June 23, 2012, from http://louisiana.baeo.org/mission.html

Bogotch, I., Miron, L., & Biesta, G. (2009). Rebuilding New Orleans Public Schools: The case of Algiers Charter School Association. In T. Townsend & I. Bogotch (Eds.), *The elusive what and the problematic how* (pp. 85–98). Rotterdam: Sense.

Broad Foundation. (2011). Education, Retrieved July 11, 2011, from http://www.broadeducation.org/investments/approach.html

Bulkley, K. E., Henig, J. R., & Levin, H. M. (Eds.). (2010). *Between public and private: Politics, governance, and the new portfolio models for urban school reform.* Cambridge, MA: Harvard Education Press.

Buras, K. L. (2010). *Pedagogy, policy, and the privatized city: Stories of dispossession and defiance from New Orleans.* New York, NY: Teachers College Press.

Chubb, J. E., & Moe, T. M. (1990). *Politics, markets, and America's schools.* Washington, DC: Brookings Institution.

Cuban, L. (1990). Reforming again, again, and again. *Educational Researcher, 19*(1), 3-13.

Cuban, L., & Usdan, M. (2002). *Powerful reforms with shallow roots: Improving America's urban schools.* New York: Teachers College Press.

DeBray-Pelot, E. H., Lubienski, C. A., & Scott, J. T. (2007). The institutional landscape of interest group politics and school choice. *Peabody Journal of Education, 82*(2–3), 204–230.

Delpit, L. (1995). *Other people's children: Cultural conflict in the classroom.* New York: The New Press.

DeVore, D. E., & Logsdon, J. (1991). *Crescent city schools: Public education in New Orleans 1841–1991.* Lafayette, LA: Center for Louisiana Studies, University of Southwestern Louisiana.

Dougherty, J. (2004). *More than one struggle: The evolution of black school reform in Milwaukee.* Chapel Hill, NC: University of North Carolina Press.

Du Bois, W. E. B. (1901). The Freedmen's Bureau. Available from *The Atlantic Monthly* The Atlantic Online. Retrieved March 1, 2010, from http://www.theatlantic.com/past/docs/issues/01mar/dubois.htm

Du Bois, W. E. B. (1989). *The souls of black folk.* New York: Bantam.

Foreman, C. H. (1998). *Black America: The road to racial uplift.* Retrieved August 8, 2012, from http://www.brookings.edu/research/articles/1998/03/spring-affirmativeaction-foreman

Gay, G. (2000). *Culturally responsive teaching.* New York: Teachers College Press.

Grimberg, S. (2009). *We shall remain, episode 5: Wounded Knee.* Retrieved June 20, 2012, from http://www.pbs.org/wgbh/amex/weshallremain/files/transcripts/WeShallRemain_5_transcript.pdf

Hoffschwelle, M. S. (2008). Rosenwald schools in the southern landscape. In A. D. Nieves & L. M. Alexander (Eds.), *"We Shall Independent Be": African American place making and the struggle to claim space in the United States.* Boulder, CO: University Press of Colorado.

Khalifa, M. (2012). A Re-New-ed paradigm in successful urban school leadership: Principal as community leader. *Educational Administration Quarterly, 48,* 387–423.

King, J. E. (Ed.). (2005). *Black education: A transformative research and action agenda for the new century.* Mahwah, NJ: Erlbaum.

Klein, N. (2007). *The shock doctrine: The rise of disaster capitalism.* New York: Metropolitan Books.

Ladson-Billings, G. (1994). *The dreamkeepers: Successful teachers of African-American children.* San Francisco: Jossey-Bass.

Madjidi, K., & Restoule, J.-P. (2008). Comparative indigenous ways of knowing and learning. In K. Mundy, K. Bickmore, R. Hayhoe, M. Madden & K. Madjidi (Eds.), *Comparative and international education: Issues for teachers* (pp. 77-106). New York: Teachers College Press.

Marshall, J. (2011). Freeing schools from Washington's education overreach. In The Heritage Foundation *WebMemo No. 3214.* Washington DC: The Heritage Foundation.

Menefee-Libey, D. (2010). Neoliberal school reform in Chicago? In K. E. Bulkley, J. R. Henig, & H. M. Levin (Eds.), *Between public and private: Politics, governance and the new portfolio models for urban school reform.* Cambridge: Harvard Education Press.

Nieves, A. D., & Alexander, L. M. (2008). *"We shall independent be": African American place making and the struggle to claim space in the United States.* Boulder, CO: University Press of Colorado.

Perlstein, D. H. (2002). Minds stayed on freedom: Politics and pedagogy in the African-American freedom struggle. *American Educational Research Journal, 39*(2), 249–277.

Perlstein, D. H. (2004). *Justice, justice: School politics and the eclipse of liberalism.* New York: Peter Lang.

Perry, A. M., & Schwam-Baird, M. M. (2010). *School by school: The transformation of New Orleans public education.* Washington: Brookings Institution and Greater New Orleans Community Data Center.

Reckhow, S. (2010). Disseminating and legitimating a new approach: The role of foundations. In K. E. Bulkley, J. R. Henig, & H. M. Levin (Eds.), *Between public and private: Politics, governance, and the new portfolio models for urban school reform* (pp. 277–306). Cambridge, MA: Harvard Education Press.

Ritea, S. (2005). *Charter schools will get federal grant.* Retrieved November 5, 2005, from http://www.nola.com/education

Ritea, S. (2007, April 30). *Left behind? Some accuse New Orleans' east bank charter schools of turning away students with special needs.* Retrieved April 30, 2007, from http://www.nola.com

Robertson, C. (2012, June 22). *Louisiana illegally fired 7,500 teachers, judge says.* Retrieved June 21, 2012, from http://www.nytimes.com/2012/06/22/education/louisiana-illegally-fired-7500-teachers-judge-rules.html

Roesgen, S. (2005). New Orleans schools in uneasy condition [radio]. In A. Silverman (Producer), *All Things Considered.* USA: National Public Radio.

Rothstein, R., Jacobsen, R., & Wilder, T. (2008). *Grading education: Getting accountability right.* Washington, DC: Economic Policy Institute and Teachers College Press.

Schwartz, H. L., Hamilton, L. S., Stecher, B. M., & Steele, J. L. (2011). *Expanded measures of school performance.* Santa Monica, CA: RAND.

Simon, D. (2007, December 13). *Millions to be given to N.O. schools.* Retrieved December 13, 2007, from http://www.nola.com/education

Stone, D. (2002). *Policy paradox: The art of political decision making* (Rev. ed.). New York: Norton.

Strike, K. A. (2010). *Small schools and strong communities: A third way of school reform.* New York: Teachers College Press.

Thomas, P. L. (2011). *Poverty matters!: A Christmas miracle.* Retrieved August 6, 2012, from http://truth-out.org/index.php?option=com_k2&view=item&id=5808:poverty-matters-a-christmas-miracle

Thorndike, E., Lee. (1904). *An introduction to the theory of mental and social measurements.* Lancaster, PA: The New Era Printing Company.

Tisserand, M. (2007, August 23). The charter school flood. Retrieved June 11, 2011, from http://www.thenation.com/article/charter-school-flood

Tye, B. B. (2000). *Hard truths: Uncovering the deep structure of schooling.* New York: Teachers College Press.

Tyler, R. W. (1950). *Basic principles of curriculum and instruction.* Chicago, Illinois: University of Chicago Press.

Vanacore, A. (2011, October 26). *New York's Mayor Bloomberg joins fray over control of state school board.* Retrieved August 8, 2012, from http://www.nola.com/education/index.ssf/2011/10/new_yorks_mayor_bloomberg_join.html

Walton Family Foundation. (2011). The Walton Family Foundation. Retrieved June 11, 2011, from http://www.waltonfamilyfoundation.org/educationreform

Warren, M. R. (2005). Communities and schools: A new view of urban education reform. *Harvard Educational Review, 75*(2), 133-173.

Wells, A. S. (1991). Choice in education: Examining the evidence on equity. *Teachers College Record, 93*(1), 137-155.

|| NINE

The Role of Assessment in Empowering/ Disempowering Students in the Critical Pedagogy Classroom

David L. Bolton and John M. Elmore

In the last quarter century, many progressive educators have pinned much of their hope for grassroots reform in our nation's schools on teacher education programs that are committed to fostering the development of teacher-intellectuals that will appreciate and carry forward an agenda of social justice via liberatory education in their respective classrooms. Certainly a fundamental inspiration for this cause can be found in the work of the late Paulo Freire (1973), who made clear the critical role that teachers play in true liberatory education: "We can educate and liberate or we can mis-educate and oppress; however, true education must always lead to liberation" (p. 32).

However, in recent years, teacher education has transformed into teacher training. Much of this perspective is fueled by what has been termed the "technization" of teaching and, by extension, teacher education. In brief, from a technical perspective, teaching is reduced to nothing more than the execution of a predetermined set of skills and, following logically, teacher education is reduced to merely training students to master those teaching skills.

The goal of teacher education at West Chester University is to develop such critical, autonomous teachers. According to the West Chester Conceptual Framework,

> At WCU, we believe teacher candidates must be more than technicians. In addition to experiences designed for learning the technical skills necessary for effective teaching, our program provides experiences to prepare teachers who will "go for broke." Today's complex world demands that teachers are innovative, risk-

> taking individuals. In the words of John Dewey and Henry Giroux, we prepare teachers who are public intellectuals.

This goal of preparing public intellectuals is in contrast to the forces in play to increase "technization" of teacher education. The emphasis by NCATE and the Pennsylvania Department of Education upon the teacher training through an ever-growing list of skills leaves little room to develop critical and independent thinking. Preservice teachers are spending so much time learning teaching skills that there is precious little time left to encourage students to think about what the purpose is of those skills. The goal is simply to achieve maximum learning. Helping future teachers to think about what that learning should be is not important since they will be given the standards to teach once they get into schools.

A significant number of the skills that teachers are to master have to do with assessment. With increased emphasis on accountability in public schools, assessment has taken an almost all-powerful role. And yet, as with many aspects of teaching, increasingly the control is being taken from the teacher with regards to assessment. While standardized testing is nothing new, the increasing number of state assessments required reduces instructional time and takes control of what is to be taught away from the teacher. That fact, in and of itself, is disempowering. However, increasingly school districts are requiring teachers to use standardized classroom tests to assure that they are teaching to the standards. Much time and effort is being expended to develop elaborate systems of accountability. All this, ironically, is promoting a mechanical view of education and leading schools away from their true purpose.

Some, but not all, programs at West Chester University require courses in assessment. These courses cover a broad spectrum of assessment skills, ranging from formative assessment to interpretation of standardized test results. Lately, there has been an increased emphasis on interpretation of student data, a skill that is starting to become increasingly important. By itself, the skill would be empowering, if it were not for the fact that its importance is so that teachers can document learning to the state. The problem is that with the increase in content, there is little time to initiate discussions of the purpose of assessment for the teacher. Indeed, such a discussion may be viewed as unnecessary given the prescriptive nature of assessment in today's public school.

Since the focus of teacher education at West Chester University has shifted toward teacher training, Democracy and Education (EDF 300), the one foundations course that students take, is often where they learn critical perspectives on education. In this class, students define and examine their own philosophies and beliefs about the purpose of education in democratic society and compare, contrast, reject, and borrow from the philosophies of others. Since one of the stated goals of education at West Chester University is to create public intellectuals, it is critical that that foundations course be as empowering as possible.

Learning to think critically about assessment should be a vital part of this foundations course. If students are critically examining the purpose and content of education, that is, instruction, then students also must learn to become critical assessors of their students. They must be given the intellectual tools to refocus the debate about assessment, so that assessment is not their master but is a tool that will empower them as they teach their own students.

In Defense of Assessment

Ideally, the role of assessment in education is to empower students. Assessment provides feedback to the students, which helps them judge whether they have learned. Assessment, therefore, should let students take charge of their education by allowing for self-evaluation. They can determine what they need to learn to become competent in the subject area. And they can then find ways to make up for any deficits.

That is the ideal. But why is this often not the case? Students often have a negative view of tests, projects, and homework, rather than viewing them as opportunities to demonstrate what they know or as ways to learn how to improve. So assessment can disempower students by making them feel inadequate and generating feelings of avoidance. This sense of disempowerment is particularly startling when seen in college students. College is supposed to help students to become independent, freethinking learners. What is happening? Why are college students so assessment-adverse? Why are assessment and learning so disconnected in the minds of the students?

For one thing, students do not come to the university tabula rasa. They bring with them a long and profound history. They are a product of an educational system that purposefully or inadvertently works against a sense of empowerment. And assessment is being used as a tool for which it was not intended: to motivate students toward educational goals.

The primary purpose of assessment should be to determine what students know and are able to do. Grades are primarily measurement instruments, like thermometers, so they should be used to take the students' temperature. The problem is that taking one's temperature will not, by itself, eradicate a fever. To reduce a fever, one needs to deal with the cause. And yet much more emphasis, time, effort, and money is being placed on identifying the problems rather than designing solutions that truly empower students.

If assessment were the solution, then today's students should be excelling beyond all measure. Students today have been exposed to more assessment than at any other time in the past history of education in the United States. Students are assessed throughout their educational career, starting in assessing their readiness for kindergarten and ending with their end-of-school project, as some schools are now requiring. And mandates increase in the upper grades. Federal mandates require assessment throughout their public school time in the areas of math, reading and writing, as well as science. And this all on top of the traditional standardized testing that has been done for decades (Hebert, 2007).

Instead of helping and empowering students, assessment is negatively impacting and disempowering students. The increased use of assessment puts a lot of stress on students, particularly as the consequences of failure on them increase. With the increased emphasis on success, students are more likely to repeat grades based on test results, and assessment could ultimately deny them a high school graduation. While some students might need to be held back or should not graduate high school, the focus on tests as the criteria for doing so is putting too much emphasis on tests, which are imperfect measures of students' knowledge.

The overemphasis of tests creates a negative attitude toward those very tests. In "The Impact of Testing upon Teacher's Instructional Practices" (Hebert, 2007), teachers are interviewed about the impact of assessment on the classroom. As a result of the district benchmark tests, Kelly, a teacher, describes student attitude toward science and science assessment: "kids love doing science but hate taking science tests." She also describes the stress on the teachers and the students:

> Our entire staff expects at least 90% of our kids to pass so that our school will be exemplary. Our principal and assistant principal check with us frequently to ask which kids are struggling and who are worried about passing. . . . The kids feel the pressure, and that is so sad to me! Do they know how important the test is, or do they just want to learn and understand science concepts? My bet is on the test! (cited in Hebert, 2007, p. 157)

Is it any wonder that school districts are caught altering the students' scores to achieve higher ratings? The economic pressures to produce higher achieving students are undermining the purpose of the tests, which is to increase learning. Instead of thoughtful, hands-on instruction in the subject area, instructional time focuses on test preparation, producing a drill-and-kill environment for the students. The curriculum is narrowed, particularly as the tests approach, with items eliminated from the curriculum that are unrelated to the test (Hebert, 2007). In the lives of these students, assessment has taken on the role of a game, a hoop to jump through, without any real relationship to what they are learning. As a result, students don't see the relationship between assessment and learning. Learning can actually be fun. But assessment is a profound negative for them.

One phenomenon related to the increase in assessment has been the increase in competitiveness among students in high school. In a meta-analysis, Johnson and Johnson (1974) reported that the competitive nature of testing is the reason students have an intense fear of the testing process. In a meta-analysis of factors related to test anxiety, ego-involving high stress was related to high levels of test anxiety (Hembree, 1988), such as the type of stress created by undue competition.

As much as students may dread assessment, they see it as a key to a better life. When that perception is combined with our society's relentless drive toward economic success, the mixture can be deadly. Students are expected to get into a good college, so GPA takes on great significance. Receiving anything less than an A on a test can be devastating for some students. Because of this pressure, stress levels of students increase. As depicted in the film *Road to Nowhere* (Abelas & Congdon, 2010), such pressures result in students' attempts to take their lives.

As with any form of anxiety that can interfere with mental processes, test anxiety can interfere with learning. Moreno (2010) states that when test anxiety gets too high, it becomes debilitating. Test anxiety affects a student's attention span as well as with his or her ability to retain information. Moreno points to the need to create an accepting and noncompetitive environment in the classroom to minimize anxiety. Unless conscious efforts are made by the teacher, grades and tests will generate a sense of competition among students, which will result in test anxiety.

The current focus on assessment has also has a negative impact on students' goal orientation. Educational psychologists talk about two types of goal orientation: mastery and performance. A mastery orientation means that a student is interested in self-improvement and increased understanding of what is being learned (Moreno, 2010). For example, a student might be interested in learning a foreign language, not to get a good grade but to improve himself or herself. A student might learn a foreign language outside of a class simply to learn the language.

A performance-oriented student learns not for the sake of learning but for the recognition of the accomplishment of the goal. Such a student is not focused on obtaining a good understanding of the topic. His or her focus is to get a good grade and to be recognized for it. Students who are performance oriented will want to compare their grades with others, so tests and grades are very important (Moreno, 2010).

A student with a mastery orientation will be persistent in working toward his or her educational goals, even in the face of adversity. That is because the goal itself has value. Such a student can see the long-term benefits and is therefore not affected by short-term setbacks or by criticism of others. On the other hand, because a performance-oriented student focuses on the short-term benefits of the learning, he or she may fail to achieve the long-term goal when facing adversity or criticism.

Anderman et al. (2010) argue that goal-orientation theory is particularly important when discussing contemporary assessment issues:

> A mastery goal structure is conveyed to students when the practices in the classroom emphasize effort, improvement, and self-comparisons. Examples of such practices might include (a) being allowed to correct errors and resubmit assignments to demonstrate mastery of a concept; (b) being recognized (and not penalized) for taking on challenging academic tasks, and (c) being encouraged to work toward mastery of academic materials. A performance goal structure is conveyed to students when the practices emphasize (a) comparisons of ability (i.e., publicity showcasing the "best" test scores or project); (b) the importance of testing and grades; and (c) demonstrating one's knowledge. (p. 124)

Although using assessment in the classroom should not primarily be about motivation, it is important to acknowledge the motivating nature of assessment. Hickey and Zuiker (2005) talk about the simplistic motivation of the No Child Left Behind legislation, which emphasized a market-oriented style of competition, rather than a scientific, research-based approach to motivation. As they pointed out, this simplistic, competitive-oriented view of motivation actually reduces motivation of students, rather than spurring them to greater achievement. Therefore, the legislation that was intended to not leave any student behind may have inadvertently facilitated the leaving behind of at-risk students:

> Inarguably, this legislation ignores the wealth of research showing that such policies will actually cause at-risk students to be "left behind" by diminishing engagement and persistence. (Hickey & Zuiker, 2005, p. 278)

The results of such policies are students who are more extrinsically motivated and not internally driven.

To understand the impact of assessment, it is important to realize that the assessment policies and practices of a teacher define and enforce the power relationship between teacher and students. It is therefore critical to understand what impact these policies and practices have. If one is to truly empower students, it is important to critically consider one's assessment policies and practices.

To be truly empowering, teachers must be willing to share control with their students. This is easier said than done. For one thing, students are often not mentally or emotionally ready to use that responsibility appropriately:

> As a new professor, I gave my college students group assignments, naively thinking that they had the understanding and maturity to divide up the assignment and complete it. While some groups could, more often than not, the group could not handle the task *as a group*. Often one or two students would take responsibility and the other students would contribute something, or occasionally nothing, to the process. (D. Bolton, personal communication, June 24, 2012)

The reality was that the students in these groups were not ready to be independent learners. The same phenomenon has been observed among master's students, where one might expect to see more evenly shared responsibility. The focus of the group work is upon the grade—and getting by—and not the learning.

It appears that the educational system they have experienced has not developed this skill in these students. Classroom control is an important part of preservice education. While a teacher must always be in charge of the classroom, such control can create an overly dependent power relationship between teacher and student. With regards to assessment, teachers become the assessors and students are the assessees. Under such a relationship, students are not allowed much input into that process. If students are to develop critical consciousness, they must be given the freedom to do so in the classroom. And if students are to become self-evaluators, then they must be allowed input into their own evaluation. If students are not exposed to anything other than a traditional teacher-student power relationship, they will take that orientation with them to college and will look for and even desire such dependent power relationships.

Based on our experience, students generally don't question assessment practices at West Chester University. And if they are asked to provide input, they may not feel comfortable doing so:

> At the end of my assessment class, I give my students a case study describing a student who is doing very well on his tests, his average is a 98, yet not doing his homework. As a result, his grade is being reduced significantly. This is based upon the teacher's grading policy of requiring students to do homework, no matter what. The boy's parents are scheduled to come in for a parent-teacher conference to discuss the situation. They are upset that the boy is not getting an "A", which they feel he deserves. The students are to discuss the situation and determine what the teacher should say to the parents. The purpose of the exercise is to get students to think about homework and its purpose. When I started doing this exercise with my students, I expected them to see that homework would not be necessary for such an individual and that deducting points from his grade would not make sense. However, I am often sorely disappointed, and even shocked, about the reaction of the students as to how adamant they can be about the necessity of following the policy. Some even expressed a sense of outrage at the audacity of the parents' challenging the teacher's rule. And as much as I talked about the fact that reducing the student's grade would mean that the grade assigned would not reflect the achievement level of the student, most could not see beyond the rules. (D. Bolton, personal communication, June 24, 2012)

In our experience, because the current power structures are what students are used to, students tend to comply—even if they aren't happy with the power structure. It is easy to follow along and it gets them what they want without too much effort. In the group work, there tends to be two types of students: the doers and the followers. Both types of students appear to be manipulating the system for their own purposes. The doers take the responsibility on themselves because they think they can do it better, but that is also their style. It is easy for them and they are getting what they want. The followers really don't want to participate, so they sit back and let the doers do. It is only when they are asked that they participate, but they hope that that will never occur.

Ultimately, an important question that every teacher and professor needs to ask is, What should really be the primary focus of learning and how can that be achieved? This question affects the whole educational process, including assessment. If the primary focus of learning is the transmission of information from the teacher to learner, the purpose of the educator is to find the most efficient way of getting that knowledge into the student. This is a behavioral/empiricist and cognitive/rationalist view of knowledge (Hickey & Zuiker, 2005). According to this view, knowledge resides outside the student and students are therefore passive recipients of this knowledge. On the other hand, if learning is participatory, consisting of "meaningful participation in the practices that characterize a community" (p. 278), then students are active in the learning process. "The latter assumes that knowledge resides in the minds of individual knowers after being constructed by rational processes" (p. 278).

Each perception requires different forms of assessment. If the learner is a passive recipient, then objective forms of assessment are appropriate. However, if the learner is to be active, then alternative forms of assessment are needed. These alternative forms of assessment engage students in the learning process and are, therefore, empowering:

> What I have found is that students often do not retain many of the details that are taught in class. Last week, for example, I was advising a former student who was going to start a master's program. I recognized the student and she recognized me. But she couldn't even tell me which class I had taught. So how was she going to retain all the individual details I had taught her if she couldn't remember my name. But what she had retained was her attitude toward learning at West Chester University. I and others had created a positive attitude, which meant she was willing to come back for more learning.
>
> What I try to focus upon in my assessment class are attitudes toward assessment. I do that by telling stories about tests and other forms of assessment from my life and the lives of others I've met—often former students who have shared. But I also do it by modeling good assessment and instructional practices. If I were to try to cram information down their throats because I've been told that these things are important, then I undermine what I am doing when I try to teach them attitudes and concepts. In my book, if I had to compromise, I feel it is more important to teach less and model more. For it is attitudes that are most important. (D. Bolton, personal communication, June 24, 2012)

If students are simply assessees, then students are being disempowered. The control over their education is being exercised by others, rather than the students. As a result, others are constantly telling them if they have succeeded or not and they are dependent upon others for their assessment. Even if colleges talk about empowerment of students (see the mission statement of West Chester University's College of Education pp. 126–127), it may be that the rules and practices that are put into place belie that intention. Teacher and faculty members are constantly bombarded with rules, which may make their job of educating students more difficult. For example, our administration is constantly asking us to bump class sizes. However, larger class sizes make it impossible for teachers and faculty members to truly implement a system that is democratic and empowers the students.

Free schools represent the ultimate freedom in education. With free schools, students have the freedom to learn whatever they desire and therefore are expected to take responsibility for their education. In the free school approach, students are not assessed by faculty, so in free schools, student self-assessment is critical. Although we all self-assess, our self-assessments may be inaccurate. Assessment is a skill that needs to be learned, and students need to be provided with guidance as to how to do it.

No matter what type of school one teaches in, to truly empower students, the focus of assessment needs to shift from a teacher-guided to a student-guided assessment process. Learning to self-assess empowers students to become independent learners. Self-assessment may involve giving students control in choosing which specific learning outcomes are important and which methods should be used to assess those outcomes. It involves deciding on criteria which will be used to evaluate oneself and how that information will be used. But ultimately, it involves having the students assess and make decisions about themselves and their learning. Such practices will help them grow in understanding of the subject matter and themselves as learners:

> Self-assessment is more accurately defined as a process by which students 1) monitor and evaluate the quality of their thinking and behavior when learning and 2) identify strategies that improve their understanding and skills. That is, self-assessment occurs when students judge their own work to improve performance as they identify discrepancies between current and desired performance. (McMillan & Hearn, 2008, p. 40)

So how can teachers and professors facilitate self-assessment? Portfolios are one method. Assessment is about gathering evidence and evaluating the evidence to make decisions. Rather than let the teacher be the evidence gatherer, portfolios let the student take control of the information gathering as an evaluation process. To do so, teachers and students typically sit down prior to starting a portfolio and discuss what types of things should be used to document their progress. They also talk about how the portfolio will be evaluated. Students are typically responsible for keeping track of their tests or documents they have created into the portfolio and then selecting the artifacts that will be used to document their progress toward the learning goals. Often students are responsible for explaining how the artifacts document that they have achieved the learning outcomes. Although the teacher may evaluate the individual artifacts, the student is ultimately responsible for the artifacts as a whole.

As can be seen from the description, the student plays a major role in this process. If this is done consistently, students learn to take responsibility for their learning and become self-assessors. When done well, the use of portfolio assessment can empower students by helping them become independent, self-critical learners. This skill will help them later in life as they participate in the work world, where they will be expected to document their work and provide evidence of what they have accomplished. For example, faculty members are required to submit tenure and promotion packages, which allow other faculty members to see what they are doing and justify their applications for tenure and promotion. The same thing happens in business, where employees are evaluated for raises and promotions, or simply evaluated as to whether they can keep their job.

Performance assessments and writing assignments are typically used ways to shift the focus from the teacher to the student. A performance assessment is a form of assessment that requires the student to demonstrate something, which is generally evaluated by the teacher using a rubric. A performance, or a writing sample, often requires that students put together individual skills learned in class into a cohesive whole. It also allows students to demonstrate their knowledge and skills directly, as opposed to objective testing, which can only indirectly measure learning outcome. While objective test formats, such as multiple choice, true-false, and matching, can be used to measure certain objectives, even higher-level thinking, only by doing a performance task, such as writing, can one be assured that students can perform the outcome measure. When the tasks being performed are meaningful, performance assessments empower students by teaching them how to perform meaningful skills that can be transferred to other situations.

Much of classroom assessment focuses on summative assessment. The purpose of summative assessment is to assign grades to students. An example of a summative assessment is a test that is given at the end of a unit to determine if students have learned the material in the unit. Formative assessments, on the other hand, provide feedback to students while they are learning. Their primary purpose is not to assign a grade, but to help students to learn by pointing out areas of weakness. An example of a formative assessment might be a quiz or homework assignment given during the unit to make sure students are learning. Ideally, formative assessments should not count toward the grade at all, or should count very little. Points are typically assigned to assure that students will take it seriously.

There is significant evidence that demonstrates that formative assessment activities in the classroom increase student motivation and achievement (McMillan & Hearn, 2008). An unspoken assumption of summative assessment is that all students learn at the same pace. While some students may learn immediately after being taught, many students require repeated exposure. Formative assessment serves as practice without major negative consequences. Repeated exposure to the subject matter will increase the likelihood that students will learn. It will also increase motivation by giving them a sense that they can master the subject.

Another way of empowering students is to practice open assessment, which involves making the assessment process as transparent as possible to students:

> As a professor who teaches classes in assessment for pre-service teachers, it is particularly important for me to model good assessment practices. And I strive to make assessment as open as possible. I want to make sure that students are very clear as to my expectations for tests and projects. To do that I provide my students with clearly worded guidelines for projects. For the projects, I have built-in checkpoints where I check to see if students are on track. If it is a group project, I assign different roles to each student to assure that each student is participating in the process. While all this structure may seem controlling, it is ultimately freeing since students can relax knowing what my expectations are. Because many students have not learned to be self-motivated learners, such structure is necessary—at least initially. As students become independent learners, much more of the control should be given over to the students. (D. Bolton, personal communication, June 24, 2012)

A further way of giving control to the students is through the use of rubrics. Again, it may seem contradictory that providing a structured description of one's expectations would be empowering. One would hope they would understand what is expected based on the instruction. But that is often not the case. What a rubric does is to focus the students' attention on the essential elements of the task. By pointing to the important aspects of the task, students can then more clearly focus on them. It eliminates the need to read the teacher's mind regarding what is important. The students can then work to more clearly understand what is important. In essence, a rubric is a teaching tool that helps students learn how to distinguish what is important from what is not. This is not to say that there are no dangers in using rubrics. If rubrics are poorly designed or used mechanically, they can point to the unimportant aspects of the task, or they can narrow the focus of the assessment only to those things that can be counted. Poor rubrics can be as disempowering as good rubrics can be empowering.

Multiple-choice questions and other types of objective tests may be appropriate for some purposes, but they need to be used cautiously and effectively if one wants to empower students:

> While I do use multiple-choice tests, I am very careful about the quality of the questions. This is particularly important since I teach a course on assessment. So I need to model good assessment practices. When I teach the students about writing objective test questions, I make them aware that what they experience in other classes is not necessarily good educational practice. As a group, we talk

> about how questions can be unclear and confusing and I talk about ways of writing good multiple-choice questions, such as avoiding all-of-the-above, none-of-the-above, multi-answer questions (A and B, B and C, etc.), as well as using negatives ("not" and "except") questions. I also explain how these can very confusing to students; particularly for those with learning disabilities. In general, I talk about the fact that some professors, because of their high level of knowledge, may have a difficult time putting themselves in the shoes of the students. So they end up writing questions that require skills that are above that of a novice, i.e., a new student.
>
> Ultimately, this unit is one of the most empowering units of my class. The comment that I typically hear from the students as I teach the class is that they see these questions all the time. However, after this unit, I often have students come back and tell me that they just took a test and it was full of poorly-written questions. At that point I can see that these students have "gotten it."
>
> The power of this unit lies in the fact that students, over the course of many years of schooling, have become desensitized to their feelings of frustrations about poorly worded questions. At one point, students may have recognized that these questions were confusing. But with time, that feeling has been suppressed. That feeling is still there though. Because they are repeatedly exposed to the same types of questions, they assume that their feelings are the problem and not the test. By acknowledging the problems with tests, it empowers students to trust their gut and to become critical consumers of assessment. (D. Bolton, personal communication, June 24, 2012)

Not everyone teaches an assessment course. However, all teachers need to be critical of their own tests, recognize that their assessments are fallible, and invite students to a discussion about the assessments they use. One way of doing that is to show the students the item statistics for a test after it has been given. In particular the p-value will be of interest to the students. The p-value shows the percent of students who have gotten a question correct. If it is low—below .65, or 65%—it may indicate a concern with the question. Looking at the percent of students who chose other options is also beneficial as it shows which other options are potential concerns. Opening up a discussion about the test may be beneficial as it can help students to learn what is being taught. Also, students are given the opportunity to exercise control over their learning:

> When I go through the test, I talk about the fact that there are three potential problems if a question is difficult: 1) Either I did not do a good job as a teacher teaching; 2) I did a poor job in creating the test; or 3) The students did not study. My discussion and questions focus on trying to determine which was the cause of the problem. If 2) or possibly 1) I will credit an option or an item. If 1) or 3), I will use the opportunity to re-teach the material.
>
> Through this process, I hope to focus the attention upon learning and away from the grades. But this process does not always go as planned. Because so few professors invite students into such a discussion, some students misunderstand what I am doing. I've been teaching a research methods course in the sociology

department. As usual, I reviewed the test with the class afterwards. I could hear gasps throughout the room as I gave credit for a question which was extremely difficult for the students, having concluded that the distinction I was trying to make was beyond what the students could grasp in the short amount of time they had to learn the subject matter. Afterwards I received the following email from a student:

> *hi mr Bolton, i just want to talk about how i fell about the 2 questions we went over in class today, i fell like it okay to give the other students extra points for getting those 2 questions wrong, but i fill like the people that get them right should have extra credit too for taking their time and choosing the right answer. i am not just speaking for myself, because i spoke to other students and i see their expression on their face, and i am not trying to disrespect you or tell you what to do. i have a 92 ON MY TEST AND I'M HAPPY WITH MY GRADE. but I JUST want to TELL YOU HOW I FEEL.* (D. Bolton, personal communication, June 24, 2012)

The purpose of the exercise is to focus on learning the subject matter, rather than on the test. It is clear from this email that the focus of this and other students was performance oriented and not mastery oriented. Indeed as one tries to empower students, one will likely come up against this performance-oriented mindset time and time again.

Part of having open assessment is to make sure students are aware of what is to be on the test. Having a discussion about what is important in preparing for the test is one way of doing this. However, that may not be enough as students struggle with the material:

> I've found that simply telling students what will be on the test is often not enough. So I provide students with sample tests, which exactly parallel the actual tests. Providing them with a sample test reduces test anxiety, which helps them to do better on the test. As with a rubric, if you want students to understand what is important, one needs to make them aware of what is to be on the test. (D. Bolton, personal communication, June 24, 2012)

These principles and practices are nothing new. In the early 1990s, outcome-based education was being touted as the way to educational reform. Outcome-based education required school districts to demonstrate how their students were achieving a set of learning outcomes. It differed from the standards-based approach here in Pennsylvania in that it gave freedom to the school districts to choose their own way of demonstrating mastery of the outcomes. Under outcome-based education, students would also be given the ability to demonstrate learning through portfolios and an end-of-school project. The approaches to assessment discussed above (portfolio assessment, performance assessment, and open assessment) were part of the outcome-based education approach to education.

Much was lost when a standards-based approach was adopted. Instead of empowering school districts to implement true educational reform, as outcome-based education was poised to do, school districts now are controlled by the myriad mandates of the federal government through the state departments of education. Instead of empowering a grassroots movement to create educational reform, schools are being disempowered by a heavy-handed federal government, which has implemented a carrot-and-stick method of motivating school districts to improve. Unfortunately this approach appears to be leaving many students behind (Hickey & Zuiker, 2005).

Critical pedagogy has much to offer with regards to assessment. It can provide the philosophical foundation upon which to create an assessment system that will produce true educational reform. However, even though critical pedagogy has the potential to impact assessment, it hasn't lived up to that potential in the past:

> Yet, as we observed at the outset, while there are some examples of a critical pedagogy affecting content and method, corresponding changes in the practice of assessment are harder to find. (Reynolds & Trehan, 2000, p. 269)

If critical pedagogy has not made an impact on practices of assessment, why is it not happening? Perhaps it is simply a lack of exposure to some of the ideas that have been promulgated among assessment reformers over the last 25 years:

> As I've talked with John and some of the other critical pedagogy faculty about what I teach in my class, I am surprised at the lack of exposure to some of the concepts that I teach and take for granted. (D. Bolton, personal communication, June 24, 2012)

Perhaps, though, the problem is systemic in nature. With as large classes as there are at West Chester University, it is difficult to provide individual attention to students. Because many students come into EDF 300 with a performance-orientation, faculty members have their work cut out for them. Since there is only one class that focuses on critical pedagogy, faculty members are working against a system that works against developing a critical consciousness in its students.

Or maybe it is simply the overwhelming amount of information that one is expected to convey to these students within EDF 300. As mentioned above, West Chester University is being forced to meet NCATE standards. The sheer number and level of these standards that need to be met in this one class is daunting. And since other faculty members often do not share the same orientation, faculty members at West Chester University feel a deep desire to make an impact on these students by sharing the wealth of information about great educational thinkers.

Ultimately, the question comes down to what should be the focus of instruction. Does one expect students to remember facts, concepts, or attitudes? Attitudes are long lasting and have a much more significant impact on behavior than facts or even concepts, for attitudes determine if students are willing to implement the concepts that one is trying to teach. Without buy-in from the students, teaching is just blowing in the wind. Without a change of attitude, one's teaching is falling on deaf ears.

Richard Lavoie in the F.A.T. City video (Peter Rosen Productions, 1989) tells a story about a man who spends a week teaching his 12-year-old son about honesty. On the first day he gives his son a long lecture on honesty. Throughout the week he uses different pedagogies to convey the value of honesty: cooperative learning, PowerPoint presentations, etc. Then on Saturday evening he and his son go to the movies. As the man is about to buy tickets for the movie, he bends down and whispers in his son's ear, "Tell him you are 10. That way, we'll save some money on the ticket." The implication of the story is that what you do has a much greater impact than what you say. In those few seconds in front of the movie theater, that boy learned more about honesty than in the hours of instruction that the father had provided during the week.

Could the same thing be happening in educational foundation classes? Ultimately, it is how we teach—and particularly assess—and not what we teach, that will have a lasting effect on our students through the formation of attitudes toward the subject matter. If those who teach critical pedagogy are truly interested in creating transformative intellectuals, then there must be a focus on using alternative assessment methods and open assessment to empower students and to make sure what is preached is practiced.

References

Abelas, V. (Producer and Co-director), & Congdon, J. (Co-director). (2010). *Road to nowhere* (Motion picture). United States: Reel Link Films.

Anderman, E. M., Anderman, L. H., Yough, M. S., & Gimbert, B. G. (2010). Value-added models of assessment: Implications for motivation and accountability. *Educational Psychologist, 45*(2), 123–137.

Freire, P. (1973). *Education for critical consciousness*. New York: Continuum.

Hebert, T. R. (2007). The impact of testing on teachers' instructional practices. *Teacher Education and Practice, 20*(2), 148–162.

Hembree, R. (1988). Correlates, causes, effects and treatment of test anxiety. *Review of Educational Research. 58*(1), 47–77.

Hickey, D. T., & Zuiker, S. J. (2005). Engaged participation: A sociocultural model of motivation with implications for educational assessment. *Educational Assessment, 10*(3), 277–305.

Johnson, D., & Johnson, R. (1974). Instructional goal structure: Cooperation, competition, or individualistic. *Review of Educational Research, 44*, 213–240.

McMillan, J. H., & Hearn, J. (2008). Student self-assessment: The key to stronger student motivation and higher achievement. *Educational Horizons, 87*(1), 40–49.

Moreno, R. (2010). *Educational psychology.* Hoboken, NJ: John Wiley.

Peter Rosen Productions (Producer). (1989). How Difficult Can This Be? F.A.T. City—A Learning Disabilities Workshop (Motion Picture). United States: PBS Video.

Reynolds, M., & Trehan, K. (2000). Assessment: A critical perspective. *Studies in Higher Education, 25*(3), 267–278.

PART TWO

De-Grading and De-Testing in a Time of High-Stakes Education Reform

||TEN

The Case Against Grades*

Alfie Kohn

"I remember the first time that a grading rubric was attached to a piece of my writing. . . . Suddenly all the joy was taken away. I was writing for a grade—I was no longer exploring for me. I want to get that back. Will I ever get that back?"
—Claire, a student (Olson, 2006)

By now enough has been written about academic assessment to fill a library, but when you stop to think about it, the whole enterprise really amounts to a straightforward two-step dance. We need to collect information about how students are doing, and then we need to share that information (along with our judgments, perhaps) with the students and their parents. Gather and report—that's pretty much it.

You say the devil is in the details? Maybe so, but I'd argue that too much attention to the particulars of implementation may be distracting us from the bigger picture—or at least from a pair of remarkable conclusions that emerge from the best theory, practice, and research on the subject: *Collecting information doesn't require tests, and sharing that information doesn't require grades.* In fact, students would be a lot better off without either of these relics from a less enlightened age.

Why tests are not a particularly useful way to assess student learning (at least the kind that matters), and what thoughtful educators do instead, are questions that must wait for another day. Here, our task is to take a hard look at the second practice, the use of letters or numbers as evaluative summaries of how well students have done, regardless of the method used to arrive at those judgments.

The Effects of Grading

Most of the criticisms of grading you'll hear today were laid out forcefully and eloquently anywhere from four to eight decades ago (Crooks, 1933; De Zouche, 1945; Kirschenbaum, Simon, & Napier, 1971; Linder, 1940; Marshall, 1968), and these early essays make for eye-opening reading. They remind us just how long it's been clear there's something wrong with what we're doing as well as just how little progress we've made in acting on that realization.

In the 1980s and '90s, educational psychologists systematically studied the effects of grades. As I've reported elsewhere (Kohn, 1999a, 1999b, 1999c), when students from elementary school to college who are led to focus on grades are compared with those who aren't, the results support three robust conclusions:

- *Grades tend to diminish students' interest in whatever they're learning.* A "grading orientation" and a "learning orientation" have been shown to be inversely related and, as far as I can tell, every study that has ever investigated the impact on intrinsic motivation of receiving grades (or instructions that emphasize the importance of getting good grades) has found a negative effect.
- *Grades create a preference for the easiest possible task.* Impress upon students that what they're doing will count toward their grade, and their response will likely be to avoid taking any unnecessary intellectual risks. They'll choose a shorter book, or a project on a familiar topic, in order to minimize the chance of doing poorly—not because they're "unmotivated" but because they're rational. They're responding to adults who, by telling them the goal is to get a good mark, have sent the message that success matters more than learning.
- *Grades tend to reduce the quality of students' thinking.* They may skim books for what they'll "need to know." They're less likely to wonder, say, "How can we be sure that's true?" than to ask "Is this going to be on the test?" In one experiment, students told they'd be graded on how well they learned a social studies lesson had more trouble understanding the main point of the text than did students who were told that no grades would be involved. Even on a measure of rote recall, the graded group remembered fewer facts a week later (Grolnick & Ryan, 1987).

Research on the effects of grading has slowed down in the last couple of decades, but the studies that are still being done reinforce the earlier findings. For example, a grade-oriented environment is associated with increased levels of cheating (Anderman & Murdock, 2007), grades (whether or not accompanied by comments) promote a fear of failure even in high-achieving students (Pulfrey et al., 2011), and the elimination of grades (in favor of a pass/fail system) produces substantial benefits with no apparent disadvantages in medical school (White & Fantone, 2010). More important, no recent research has contradicted the earlier "big three" findings, so those conclusions still stand.

Why Grading Is Inherently Problematic

> A student asked his Zen master how long it would take to reach enlightenment. "Ten years," the master said. But, the student persisted, what if he studied very hard? "Then 20 years," the master responded. Surprised, the student asked how long it would take if he worked very, *very* hard and became the most dedicated student in the Ashram. "In that case, 30 years," the master replied. His explanation: "If you have one eye on how close you are to achieving your goal, that leaves only one eye for your task."

To understand why research finds what it does about grades, we need to shift our focus from educational measurement techniques to broader psychological and pedagogical questions. The latter serve to illuminate a series of misconceived assumptions that underlie the use of grading.

Motivation: While it's true that many students, after a few years of traditional schooling, could be described as motivated by grades, what counts is the nature of their motivation. Extrinsic motivation, which includes a desire to get better grades, is not only different from, but often undermines, intrinsic motivation, a desire to learn for its own sake (Kohn, 1999a). Many assessment specialists talk about motivation as though it were a single entity—and their recommended practices just put a finer gloss on a system of rewards and punishments that leads students to chase marks and become less interested in the learning itself. If nourishing their *desire* to learn is a primary goal for us, then grading is problematic by its very nature.

Achievement: Two educational psychologists pointed out that "an overemphasis on assessment can actually undermine the pursuit of excellence" (Maehr & Midgley, 1996, p. 7). That unsettling conclusion—which holds regardless of the quality of the assessment but is particularly applicable to the use of grades—is based on these researchers' own empirical findings as well as those of many others, including Carol Dweck, Carole Ames, Ruth Butler, and John Nicholls (for a review, see Kohn, 1999b, Chapter 2). In brief: the more students are led to focus on *how well* they're doing, the less engaged they tend to be with *what* they're doing.

It follows that all assessment must be done carefully and sparingly lest students become so concerned about their achievement (how good they are at doing something—or, worse, how their performance compares to others') that they're no longer thinking about the learning itself. Even a well-meaning teacher may produce a roomful of children who are so busy monitoring their own reading skills that they're no longer excited by the stories they're reading. Assessment consultants worry that grades may not accurately reflect student performance; educational psychologists worry because grades fix students' attention *on* their performance.

Quantification: When people ask me, a bit defensively, if it isn't important to measure how well students are learning (or teachers are teaching), I invite them to rethink their choice of verb. There is certainly value in *assessing* the quality of learning and teaching, but that doesn't mean it's always necessary, or even possible, to *measure* those things—that is, to turn them into numbers. Indeed, "measurable outcomes may be the least significant results of learning" (McNeil, 1986, p. xviii)—a realization that offers a refreshing counterpoint to today's corporate-style "school reform" and its preoccupation with data.

To talk about what happens in classrooms, let alone in children's heads, as moving forward or backward in specifiable degrees, is not only simplistic because it fails to capture much of what is going on, but also destructive because it may change what is going on for the worse. Once we're compelled to focus only on what can be reduced to numbers, such as how many grammatical errors are present in a composition or how many mathematical algorithms have been committed to memory, thinking has been severely compromised. And that is exactly what happens when we try to fit learning into a four- or five- or (heaven help us) 100-point scale.

Curriculum: "One can have the best assessment imaginable," Howard Gardner (1991, p. 254) observed, "but unless the accompanying curriculum is of quality, the assessment has no use." Some people in the field are candid about their relativism, offering to help align your assessment to whatever your goals or curriculum may be. The result is that teachers may become more adept at measuring how well students have mastered a collection of facts and skills whose value is questionable—and never questioned. "If it's not worth teaching, it's not worth teaching well," as Eliot Eisner (2001, p. 370) likes to say. Nor, we might add, is it worth assessing accurately.

Portfolios, for example, can be constructive if they replace grades rather than being used to *yield* them. They offer a way to thoughtfully gather a variety of meaningful examples of learning for the students to review. But what's the point, "if instruction is dominated by worksheets so that every portfolio looks the same"? (Neill et al., 1995, p. 4). Conversely, one sometimes finds a mismatch between more thoughtful forms of pedagogy—say, a workshop approach to teaching writing—and a depressingly standardized assessment tool like rubrics (Wilson, 2006).

Improving Grading: A Fool's Errand?

"I had been advocating standards-based grading, which is a very important movement in its own right, but it took a push from some great educators to make me realize that if I wanted to focus my assessment around authentic feed-

> back, then I should just abandon grades altogether."—New Jersey middle school teacher Jason Bedell (2010)

Much of what is prescribed in the name of "assessing for learning" (and, for that matter, "formative assessment") leaves me uneasy: The recommended practices often seem prefabricated and mechanistic; the imperatives of data collection seem to upstage the children themselves and the goal of helping them becomes more enthusiastic about what they're doing. Still, if it's done only occasionally and with humility, I think it's possible to assess for learning. But *grading* for learning is, to paraphrase a 1960's-era slogan, rather like bombing for peace. Rating and ranking students (and their efforts to figure things out) is inherently counterproductive.

If I'm right—more to the point, if all the research to which I've referred is taken seriously—then the absence of grades is a necessary, though not sufficient, condition for promoting deep thinking and a desire to engage in it. It's worth lingering on this proposition in light of a variety of efforts to sell us formulas to improve our grading techniques, none of which address the problems of grading, per se.

- It's not enough to replace letters or numbers with labels ("exceeds expectations," "meets expectations," and so on). If you're sorting students into four or five piles, you're still grading them. Rubrics typically include numbers as well as labels, which is only one of several reasons they merit our skepticism (Kohn, 2006; Wilson, 2006).
- It's not enough to tell students in advance exactly what's expected of them. "When school is seen as a test, rather than an adventure in ideas," teachers may persuade themselves they're being fair "if they specify, in list-like fashion, exactly what must be learned to gain a satisfactory grade . . . [but] such schooling is unfair in the wider sense that it prepares students to pass other people's tests without strengthening their capacity to set their own assignments in collaboration with their fellows" (Nicholls & Hazzard, 1993, p. 77).
- It's not enough to disseminate grades more efficiently—for example, by posting them on-line. There is a growing technology, as the late Gerald Bracey once remarked, "that permits us to do in nanoseconds things that we shouldn't be doing at all" (quoted in Matthews, 2006). In fact, posting grades on-line is a significant step backward because it enhances the salience of those grades and therefore their destructive effects on learning.
- It's not enough to add narrative reports. "When comments and grades coexist, the comments are written to justify the grade" (Wilson, 2009, p. 60). Teachers report that students, for their part, often just turn to the grade and ignore the comment, but "when there's only a comment, they read it," says high school English teacher Jim Drier. Moreover, research suggests

that the harmful impact of grades on creativity is no less (and possibly even more) potent when a narrative accompanies them. Narratives are helpful only in the absence of grades (Butler, 1988; Pulfrey et al., 2011).

- It's not enough to use "standards-based" grading. That phrase may suggest any number of things—for example, more consistency, or a reliance on more elaborate formulas, in determining grades; greater specificity about what each grade signifies; or an increase in the number of tasks or skills that are graded. At best, these prescriptions do nothing to address the fundamental problems with grading. At worst, they exacerbate those problems. In addition to the simplistic premise that it's always good to have more data, we find a penchant shared by the behaviorists of yesteryear that learning can and should be broken down into its components, each to be evaluated separately. And more frequent temperature-taking produces exactly the kind of disproportionate attention to performance (at the expense of learning) that researchers have found to be so counterproductive.

The term "standards-based" is sometimes intended just to mean that grading is aligned with a given set of objectives, in which case our first response should be to inquire into the value of those objectives (as well as the extent to which students were invited to help formulate them). If grades are based on state standards, there's particular reason to be concerned since those standards are often too specific, age-inappropriate, superficial, and standardized by definition. In my experience, the best teachers tend to be skeptical about aligning their teaching to a list imposed by distant authorities, or using that list as a basis for assessing how well their students are thinking.

Finally, "standards-based" may refer to something similar to criterion-based testing, where the idea is to avoid grading students on a curve. (Even some teachers who don't do so explicitly nevertheless act as though grades ought to fall into something close to a normal distribution, with only a few students receiving A's. But this pattern is not a fact of life, nor is it a sign of admirable "rigor" on the teacher's part. Rather, "it is a symbol of failure—failure to teach well, failure to test well, and failure to have any influence at all on the intellectual lives of students" [Milton, Pollio, & Eison, 1986].) This surely represents an improvement over a system in which the number of top marks is made artificially scarce and students are set against one another. But here we've peeled back the outer skin of the onion (competition) only to reveal more noxious layers beneath: extrinsic motivation, numerical ratings, the tendency to promote achievement at the expense of learning.

If we begin with a desire to assess more often, or to produce more data, or to improve the consistency of our grading, then certain prescriptions will follow. If, however, our point of departure isn't mostly about the grading, but about our desire for students to understand ideas from the inside out, or to get a kick out of playing with words and numbers, or to be in charge of their own learning, then we will likely end up elsewhere. We may come to see grading as a huge, noisy, fuel-guzzling, smoke-belching machine that constantly requires repairs and new parts, when what we should be doing is pulling the plug.

Deleting—or at Least Diluting—Grades

"Like it or not, grading is here to stay" is a statement no responsible educator would ever offer as an excuse for inaction. What matters is whether a given practice is in the best interest of students. If it isn't, then our obligation is to work for its elimination and, in the meantime, do what we can to minimize its impact.

Replacing letter and number grades with narrative assessments or conferences—qualitative summaries of student progress offered in writing or as part of a conversation—is not a utopian fantasy. It has already been done successfully in many elementary and middle schools and even in some high schools, both public and private (Kohn, 1999c). It's important not only to realize that such schools exist but to investigate *why* they've eliminated grades, how they've managed to do so (hint: the process can be gradual), and what benefits they have realized.

Naturally objections will be raised to this—or any—significant policy change, but once students and their parents have been shown the relevant research, reassured about their concerns, and invited to participate in constructing alternative forms of assessment, the abolition of grades proves to be not only realistic but an enormous improvement over the status quo. Sometimes it's only after grading has ended that we realize just how harmful it's been.

To address one common fear, the graduates of grade-free high schools are indeed accepted by selective private colleges and large public universities—on the basis of narrative reports and detailed descriptions of the curriculum (as well as recommendations, essays, and interviews), which collectively offer a fuller picture of the applicant than does a grade-point average. Moreover, these schools point out that their students are often more motivated and proficient learners, thus better prepared for college, than their counterparts at traditional schools who have been preoccupied with grades.

In any case, college admission is surely no bar to eliminating grades in elementary and middle schools because colleges are largely indifferent to what students have done before high school. That leaves proponents of grades for younger children to fall back on some version of an argument I call "BGUTI": Better Get Used To It (Kohn, 2005). The claim here is that we should do unpleasant and unnecessary things to children now in order to prepare them for the fact that just such things will be done to them later. This justification is exactly as absurd as it sounds, yet it continues to drive education policy.

Even when administrators aren't ready to abandon traditional report cards, individual teachers can help to rescue learning in their own classrooms with a two-pronged strategy to "neuter grades," as one teacher described it. First, they can stop putting letter or number grades on individual assignments and instead offer only qualitative feedback. Report cards are bad enough, but the destructive effects reported by researchers (on interest in learning, preference for challenge, and quality of thinking) are compounded when students are rated on what they do in school day after day. Teachers can mitigate considerable harm by replacing grades with authentic assessments; moreover, as we've seen, any feedback they may already offer becomes much more useful in the absence of letter or number ratings.

Second, although teachers may be required to submit a final grade, there's no requirement for them to decide unilaterally what that grade will be. Thus, students can be invited to participate in that process either as a negotiation (such that the teacher has the final say) or by simply permitting students to grade themselves. If people find that idea alarming, it's probably because they realize it creates a more democratic classroom, one in which teachers must create a pedagogy and a curriculum that will truly engage students rather than allow teachers to coerce them into doing whatever they're told. In fact, negative reactions to this proposal ("It's unrealistic!") point up how grades function as a mechanism for controlling students rather than as a necessary or constructive way to report information about their performance.

I spoke recently to several middle and high school teachers who have de-graded their classes. Jeff Robbins, who has taught eighth-grade science in New Jersey for 15 years, concedes that "life was easier with grades" because they take so much less time than meaningful assessment. That efficiency came at a huge cost, though, he noticed: Kids were stressed out and also preferred to avoid intellectual risks. "They'll take an easier assignment that will guarantee the A."

Initially Robbins announced that any project or test could be improved and resubmitted for a higher grade. Unfortunately, that failed to address the underlying problem, and he eventually realized he had to stop grading entirely. Now, he offers comments to all of his 125 students "about what they're doing and what they need to improve on" and makes abbreviated notes in his grade book. At the end of the term, over a period of about a week, he grabs each student for a conversation at some point—"because the system isn't designed to allow kids this kind of feedback"—asking "what did you learn, how did you learn it. Only at the very end of the conversation [do] I ask what grade will reflect it . . . and we'll collectively arrive at something." Like many other teachers I've spoken to over the years, Robbins says he almost always accepts students' suggestions because they typically pick the same grade that he would have.

Jim Drier, an English teacher at Mundelein High School in Illinois who has about 90 students ranging "from at-risk to A.P.," was relieved to find that it "really doesn't take that long" to write at least a brief note on students' assignments—"a reaction to what they did and some advice on how they might improve." But he never gives them "a number or grade on anything they do. The things that grades make kids do are heartbreaking for an educator": arguing with teachers, fighting with parents, cheating, memorizing facts just for a test and then forgetting them. "This is not why I became a teacher."

Without grades, "I think my relationships with students are better," Drier says. "Their writing improves more quickly and the things they learn stay with them longer. I've had lots of kids tell me it's changed their attitude about coming to school." He expected resistance from parents but says that in three years only one parent has objected, and it may help that he sends a letter home to explain exactly what he's doing and why. Now two of his colleagues are joining him in eliminating grades.

Drier's final grades are based on students' written self-assessments, which, in turn, are based on their review of items in their portfolios. He meets with about three-quarters of them twice a term, in most cases briefly, to assess their performance and, if necessary (although it rarely happens), to discuss a concern about the grade they've suggested. Asked how he manages without a grade book full of letters or numbers, Drier replies, "If I spend 18 weeks with them, I have a pretty good idea what their writing and reasoning ability is."

A key element of authentic assessment for these and other teachers is the opportunity for students to help design the assessment and reflect on its purposes—individually and as a class. Notice how different this is from the more common variant of self-assessment in which students merely monitor their progress toward the teacher's (or legislature's) goals and in which they must reduce their learning to numerical ratings with grade-like rubrics.

Points of overlap as well as divergence emerge from the testimonies of such teachers, some of which have been collected by Joe Bower (n.d.), an educator in Red Deer, Alberta. Some teachers, for example, *evaluate* their students' performance (in qualitative terms, of course), but others believe it's more constructive to offer only *feedback*—which is to say, information. On the latter view, "the alternative to grades is description" and "the starting point for description is a plain sheet of paper, not a form which leads and homogenizes description" (Marshall, 1968, pp. 131, 143).

Teachers also report a variety of reactions to de-grading not only from colleagues and administrators but also from the students themselves. John Spencer (2010), an Arizona middle school teacher, concedes that "many of the 'high performing' students were angry at first. They saw it as unfair. They viewed school as work and their peers as competitors. . . . Yet, over time they switch and they calm down. They end up learning more once they aren't feeling the pressure" from grades.

Indeed, research suggests that the common tendency of students to focus on grades doesn't reflect an innate predilection or a "learning style" to be accommodated; rather, it's due to having been led for years to work for grades. In one study (Butler, 1988), some students were encouraged to think about how well they performed at a creative task while others were just invited to be imaginative. Each student was then taken to a room that contained a pile of pictures that other people had drawn in response to the same instructions. It also contained some information that told them how to figure out their "creativity score." Sure enough, the children who were told to think about their performance now wanted to know how they had done relative to their peers; those who had been allowed to become immersed in the task were more interested in seeing *what* their peers had done.

Grades don't prepare children for the "real world"—unless one has in mind a world where interest in learning and quality of thinking are unimportant. Nor are grades a necessary part of schooling, any more than paddling or taking extended dictation could be described that way. Still, it takes courage to do right by kids in an era when the quantitative matters more than the qualitative, when meeting (someone else's) standards counts for more than exploring ideas, and when anything "rigorous" is automatically assumed to be valuable. We have to be willing to challenge the conventional wisdom, which in this case means asking not how to improve grades but how to jettison them once and for all.

References

Anderman, E. M., & Murdock, T. B., eds. (2007). *Psychology of academic cheating.* Burlington, MA: Elsevier Academic Press.

Bedell, J. (2010, July 23). Grading moratorium: Jason Bedell. for the love of learning [Web log]. Retrieved from http://www.joebower.org/2010/07/grading-moratorium-jason-bedell.html

Bower, J. (2010, March 28). Detoxing students from grade-use. for the love of learning [Web log]. Retrieved from http://www.joebower.org/2010/03/detoxing-students-from-grade-use.html

Bower, J. (n.d.). Grading moratorium. for the love of learning [Web log]. Retrieved from www.joebower.org/p/grading-moratorium.html

Butler, R. (1988). Enhancing and undermining intrinsic motivation: The effects of task-involving and ego-involving evaluation on interest and performance. *British Journal of Educational Psychology, 58*, 1–14.

Butler, R. (1992). What young people want to know when: Effects of mastery and ability goals on interest in different kinds of social comparisons. *Journal of Personality and Social Psychology*, 62, 934–43.

Crooks, A. D. (1933). Marks and marking systems: A digest. *Journal of Educational Research, 27*(4), 259–72.

De Zouche, D. (1945). "The wound *is* mortal": Marks, honors, unsound activities. *The Clearing House, 19*(6), 339–44.

Eisner, E. W. (2001, January). What does it mean to say a school is doing well? *Phi Delta Kappan*, 367–72.

Gardner, H. (1991). *The unschooled mind: How children think and how schools should teach.* New York: Basic Books.

Grolnick, W. S., & Ryan, R. M. (1987). Autonomy in children's learning: An experimental and individual difference investigation. *Journal of Personality and Social Psychology, 52*, 890–98.

Kirschenbaum, H., Simon, S. B., & Napier, R. W. (1971). *Wad-ja-get?: The grading game in American education*. New York: Hart.
Kohn, A. (1999a). *Punished by rewards: The trouble with gold stars, incentive plans, A's, praise, and other bribes*. Rev. ed. Boston: Houghton Mifflin.
Kohn, A. (1999b). *The schools our children deserve: Moving beyond traditional classrooms and "tougher standards."* Boston: Houghton Mifflin.
Kohn, A. (1999c, March). From degrading to de-grading. *High School Magazine*, 38–43.
Kohn, A. (2001, September 26). Beware of the standards, not just the tests. *Education Week*, 52, 38.
Kohn, A. (2005, September 7). Getting hit on the head lessons. *Education Week*, pp. 52, 46–47.
Kohn, A. (2006). The trouble with rubrics. *English Journal, 95*(4), 12–15.
Linder, I. H. (1940, July). Is there a substitute for teachers' grades? *School Board Journal*, 25, 26, 79.
Maehr, M. L., & Midgley, C. (1996). *Transforming school cultures*. Boulder, CO: Westview.
Marshall, M. S. (1968). *Teaching without grades*. Corvallis, OR: Oregon State University Press.
Matthews, J. (2006, November 14). Just whose idea was all this testing? *Washington Post*.
McNeil, L. M. (1986). *Contradictions of control: School structure and school knowledge*. New York: Routledge & Kegan Paul.
Milton, O., Pollio, H. R., & Eison, J. A. (1986). *Making sense of college grades*. San Francisco: Jossey-Bass.
Neill, M., Bursh, P., Schaeffer, B., Thall, C., Yohe, M., & Zappardino, P. (1995). *Implementing performance assessments: A guide to classroom, school, and system reform*. Cambridge, MA: FairTest.
Nicholls, J. G., & Hazzard, S. P. (1993). *Education as adventure: Lessons from the second grade*. New York: Teachers College Press.
Olson, K. (2006, November 8). The wounds of schooling. *Education Week*, 28–29.
Pulfrey, C., Buch, C., & Butera, F. (2011). Why grades engender performance-avoidance goals: The mediating role of autonomous motivation. *Journal of Educational Psychology, 103*, 683–700.
Spencer, J. (2010, July). Grading moratorium: John Spencer. for the love of learning [Web log]. Retrieved from http://www.joebower.org/2010/07/grading-moratorium-john-spencer.html
White, C. B., & Fantone, J. C. (2010). Pass-fail grading: Laying the foundation for self-regulated learning. *Advances in Health Science Education, 15*, 469–77.
Wilson, M. (2006). *Rethinking rubrics in writing assessment*. Portsmouth, NH: Heinemann.
Wilson, M. (2009, November). Responsive writing assessment. *Educational Leadership*, 58–62.

|| ELEVEN

Reduced to Numbers*

From Concealing to Revealing Learning

Joe Bower

> "A mark or a grade is an inadequate report of an inaccurate judgment by a biased and variable judge of the extent to which a student has attained an indefinite amount of material."—Paul Dressel (1957)

I am not the same teacher I used to be. When I started, I was focused on power and control. I assigned loads of homework, dished out huge penalties for late assignments, assigned punishments for rule-breaking behavior, and averaged marks to determine the students' final grade. I did some of these things because I was trained to do so in university. However, most of these teaching strategies were being done mindlessly, and, for the most part, I was simply teaching the way I was taught.

It took only six years before I wanted to quit teaching. I had become increasingly unhappy with my teaching and my students' learning. I was tired of laboring through hours and hours of marking, and I hated nagging kids to complete their homework. Instead of students asking "What is this question worth?" I wanted them to actually get excited about the content. I wanted change, and I came close to thinking that change required me to leave the profession.

Instead of pulling the plug on what could have been a short teaching career, I started to question the traditional pedagogy that I had so mindlessly adopted. I began asking questions that would challenge the status quo. Many professional development conferences provide teachers with opportunity to ask questions such as "How do I mark better?" or "How do I get my students to do their homework?" At first glance these look like challenging and provocative questions, but they are still questions that promote more of the same. Far more powerful questions are "Why do I mark?" or "Why do I assign homework?" Investigating the motives for our actions, rather than merely examining our methods of implementation, is a better use of our time, particularly if the subject in question is a belief or habit that we've come to accept as a given truth. I have come to see that

> [t]here is a time to admire the grace and persuasive power of an influential idea, and there is a time to fear its hold over us. The time to worry is when the idea is so widely shared that we no longer even notice it, when it is so deeply rooted that it feels to us like plain common sense. At the point when objections are not answered anymore because they are no longer even raised, we are not in control: we do not have the idea; it has us. (Kohn, 1999a, p. 1)

For too long, I was letting schooling get in the way of my teaching and too many of my teaching practices were based on pedagogy that was at best unhelpful and at worst harmful to my long-term goals. Through critical questioning and extensive research, I came to the conclusion that my pedagogy had to revolve around one priority: learning. If there were things that worked to sabotage learning, then it was my professional responsibility to remove them.

Since 2006, I have worked to identify and remove things like grading that traditional school has done for so long. And when I share this with others, I receive mixed responses. Some listen intently, nodding their heads in agreement, as if deep down they have always sensed something wrong with what Seymour Papert (1988) described as School with a capital S, which is a place that he explains as having a bureaucracy that has its own interests and is not open to what is in the best interest of the children. Unfortunately, when most people close their eyes and think of their Schooling, many have experienced no other kind of School than the one with a capital S. Some listen in shock and awe at how school could even function without such things as grading. The people who have a hard time comprehending how children could learn without extrinsic manipulators concern me the most. They are so invested in traditional schooling that they have never questioned its foundation. Unfortunately, some have a distrustful view of the nature of children; they believe that without grading there would be nothing to stop children from running amok.

The Day I Abolished Grading

Old School is not a place; it is a state of mind that thinks very little of the mind. I remember the day in November 2006 that I chose to no longer subscribe to Old School. I remember searching the Web for alternatives to grading and finding an article titled "The Costs of Overemphasizing Achievement" (Kohn, 1999b). It was the first Alfie Kohn article I had ever read, but it would be far from the last, and it proved to be the pedagogical pill that I had been looking for to cure my ailments for grading. I remember devouring the article and returning to my grade 8 classroom the next day with a sense of revitalized urgency that I could not wait to share with my students.

That year my teaching assignment included two classes of about 30 grade 8 students whom I taught language arts and science. In an effort to integrate the two subjects, I had assigned my students to write an essay on the particle model of matter. As far as they knew, I should have been grading their papers, but I was about to blow their minds. I walked into class and announced that I had decided not to grade their essays. I was beaming with excitement; they were not. Suddenly, the air beneath my wings had disappeared. My excitement was lost on them, and I was disheartened. But what happened next both appalled and enlightened me. I stood there at the front of the class and heard what sounded like all 30 of them yell in unison, "You mean we did this all for nothing?" Initially I felt like I had been kicked in the gut, but then I felt like the Grinch when his heart grew three sizes that day.

Their disgust was all the proof I needed to tell me I was on to something. They had completed their essays because they expected a grade, and they figured I had better keep my end of the bargain. They did not care about the particle model of matter. They did not care about their essays' sentence structure or paragraphing. There was no love for learning. It was a game that I was perpetuating, and I was done perpetuating it. I remember laughing to myself, thinking, *Is this all a facade? Why are we here?* I had to slap myself before these existential questions went too far. I spent the next few months sharing, explaining, collaborating, detailing, showing, and working with my students on how I came to the conclusion that grading had to go. A few brainiacs did not agree. Some thought I was nuts. Most cared. All listened.

I took a risk that day. My course outline had suddenly become null and void. My students had become formative assessment guinea pigs. So how did I survive? Well, at the time, I cannot even tell you that I was all that well read on the subject of real learning, formative assessment, or abolishing grading. I was inexperienced and more than a little indulgent. I survived because people trusted and supported me. My administration, students, and their parents trusted me. I was afforded enough room to work that I could shape my teaching around the way I aspired to teach. As an educator, it was the day I reinvented how I taught and why my students learned.

No Good Reason to Grade

Once I was able to move past simply asking "How do I grade better?" I realized that I needed to focus more on "Why do I grade at all?" There is a lot of common sense around why we grade students, but if you look closely, the conventional wisdom does not make all that much sense and is unfortunately all too common.

- We grade to motivate. Conventional wisdom tells us we grade students to artificially induce their extrinsic motivation to strive for the reward of a high grade, or to avoid the punishment of a low grade. Either way, it is the carrot or the stick that is the driving force. The problem here is that "we need to stop asking 'How motivated are my students?' and start asking 'How are my students motivated?'" (Brandt, 1995. para. 40). Motivation is not a single entity that you either have a lot or little of (Deci, 1996). There are two kinds: intrinsic and extrinsic. If you are intrinsically motivated, then you are doing something for its own sake; if you are extrinsically motivated, you are driven to do something, or not do something, based on a reward or punishment that may be waiting for you. But that is not even the interesting part—the real catch here is that these two kinds of motivation tend to be inversely related (Kohn, 1999a). When you grow students' extrinsic motivation by bribing them with high grades or threatening them with low grades, you run the risk of growing their extrinsic motivation while their intrinsic love for what you want them to learn shrivels. Grades can only ever gain short-term compliance from students when what we really desire is their authentic engagement. If we give grades and our students are uninterested or disengaged, might it be because they are searching for a more intrinsically motivating reason to learn? I suppose we could use grades to artificially induce children to learn, but I would hope we could help them find a real reason to do so. Needless to say, playing on children's extrinsic motivation to learn is, at best, a questionable practice.
- We grade to rank and sort. Grades allow us to place students nicely on an artificial hierarchy so that we can keep detailed records of who is at the top and who is at the bottom. When we track students like this, we encourage them to see learning as a competitive endeavor where their peers are nothing more than obstacles to be conquered to guarantee one's own success, and by definition, when students compete against one another, they cooperate less. The problem here is that competition is for the strong, and public education is for everyone. We need to spend far less time caring about who is beating whom and care far more about helping all children learn. Sadly, focusing our limited time, effort, and resources on ranking and sorting children is often easier than ensuring that they all learn. Some countries such as Finland normally avoid grade-based assessments for elementary students because they want to avoid having their children ranked and sorted (Sahlberg, 2011).
- We grade to provide feedback. By definition, a grade is any attempt to reduce learning to a symbol. It is important to note that 75%, B-, and

"proficient" have distinctions without a difference. A grade by any other name is still a grade. Grades provide students and parents with an idea where they stand in comparison to their peers, without giving them the information they need to figure out what they have done well or how they can improve. According to Butler (1988), grades are a primitive form of feedback that grows students' egos and head size without ever encouraging them to grow intellectually. Even if teachers somehow developed the "perfect" grading system that reliably ranked and sorted children with impeccable validity, placing the proper label on each child as if they were a slab of meat ready to go to market, it is likely we could give the system everything it could ever want while providing the children nothing they really need. There is also an argument to be made that learning cannot be measured but it can be observed and described (Lucido, 2012). Because "measurable outcomes may be the least significant results of learning" (Linda McNeil as cited in Kohn, 1999e, p. 75), there is no need for us to labor over reducing student learning to a grade—it is simply not why I became a teacher nor is it a good use of my limited time and effort. In other words, the bane of reducing learning to a number is that it inevitably overvalues what can be quantified and undervalues what cannot. As for parents, the best feedback they can receive about their children's learning is to observe their children learning—no reductionist data are required. It might be argued that the best kinds of information about children's learning might never fit on the refrigerator.

- We grade to prepare children for future grading. We have a dangerous preoccupation with preparation. College does not begin in kindergarten; kindergarten begins in kindergarten (Bonawitz et al., 2011). When school confuses harder with better and becomes a race to do everything earlier and faster, school becomes developmentally inappropriate (Kohn, 1999c). Holding K–12 hostage to postsecondary's archaic grading practices is but one more way to stifle innovation.

What Replaces Grades

I see the case against grades as having two parts. One is to make the case for why we should not be grading at all, and the other is to make a case for what we should be doing in place of grading. Because the primary purpose of assessment is to improve student learning, I have come to live by what Alfie Kohn has identified as Bruner's Law, which is to say that we should try and create an environment where students can "experience success and failure not as reward and punishment, but as information" (as cited in Kohn, 1999e, p. 191). In a sentence, Bruner outlines for us both what we should and should not be doing when we assess. We often seriously overestimate the effectiveness of judgment and evaluation as a precondition of learning, and because grades can only ever be experienced as a reward or punishment, they have to go (Kohn, 1999e). Bruner also challenges us to tailor our assessments in a way that the learner will inevitably experience as information. We want to avoid invoking an emotional response and ensure that the feedback causes the student to think and engage in reflective learning (Jones & Wiliam, 2008).

The Latin root for assessment is *assidere,* which means "to sit beside." This is precisely why there is no substitute for what a teacher observes when working with students while they are still learning. When I sit beside my students, I often guide my feedback in a three-step process where I (a) share what I see and hear, (b) make suggestions, and (c) ask reflective questions.

- What I see. This step is all about observing students while they are still learning. This is not about what I like or do not like, so it is critically important that I refrain from judging. (I see you have periods at the ends of your sentences; I see you included information that compared and contrasted viscosity and density.)
- Suggestions. I tend to frame my suggestions by prefacing them with either "Consider next time to . . ." or "Continue next time to . . ." (Consider next time to include punctuation at the end of all your sentences; Continue to make your thinking clear about how viscosity and density are not the same thing.)
- Questions. I try to ask insightful questions that invoke thoughtfulness from students. I stay away from questions that have one right answer and questions that could be answered with one word. The purpose here is to encourage students to reflect on their learning. (If you were to write a couple more sentences, how could you use punctuation marks other than periods? How could you compare and contrast viscosity and density to another concept?)

It is tempting to adopt a more-the-merrier attitude toward providing students with both a comment and a grade, but the research shows that the presence of a grade (with or without a comment) is responsible for lower levels of motivation, a loss of interest for learning, and a preference for easier tasks (Black et al., 2004; Butler, 1988; Pulfrey, Buch, & Butera, 2011). Unfortunately, the positive benefits of the formative comment are overshadowed by the negative effects of the grade. In other words, the power of formative comments requires the absence of grades.

Wooden Meets Bruner

I often find it helpful to share real-world examples where this kind of teaching and coaching exist. The following is my favorite.

John Wooden is considered by *The Sporting News* to be the greatest coach in American sports. To say that Wooden was a good coach is like saying Einstein was good at science. He coached basketball at UCLA for 12 years where he won 10 NCAA titles (including 7 in a row). In those 12 years he won 664 games out of 826 opportunities (that is a .804 winning percentage), and during that time the UCLA Bruins ran the table with an 88-game winning streak that took place over almost three years (four perfect 30-0 seasons).

If you were to go back in time and visit one of John Wooden's practices, you might expect to see him as the sage on the stage, providing his students with all of the knowledge they would need in his chalk-and-talk lectures. You might assume you would see him praising the hard workers and punishing the lollygaggers. However, if you were to make these assumptions, you would be wrong. When Ron Gallimore and Roland Tharp, two education psychologists, attended John Wooden's practices for the first time (while conducting a study of Wooden), they were shocked to see almost none of the above. When they observed Wooden, they were at first quite perplexed because it appeared like he hardly coached at all. "We thought we knew what coaching was," Gallimore said (as cited in Coyle, 2009, p. 168).

But on closer inspection, they found that he used his time to observe a lot and make short and quick comments to his athletes while they were actually performing. Wooden did not give speeches. He did not dole out punishments, nor did he hand out rewards and praise. In fact, when Gallimore and Roland actually recorded and categorized 2,326 of Coach Wooden's acts of teaching, they found that only 6.6% were acts of disapproval while 6.9% were acts of praise (Coyle, 2009). That means the majority of his interactions with his athletes were judgment-free statements of information.

If there was ever a coach who sports fans could agree should be given license to pass judgment, it would likely be John Wooden. But when given the chance, even Wooden preferred to reserve judgment. I am not sure if Wooden and Bruner ever sat down for a drink to discuss pedagogy, but it would appear that Wooden was drinking what Bruner was pouring.

Detoxing Students from Grade Use

I have come to see grades as school's drug of choice, and we are all addicted. Since 2006, I have tailored my assessment practices toward focusing less on grading and more on learning. Here are the six steps that summarize my experience with students and their withdrawal from grade use.

Stage 1: Intervention

At the beginning of the school year, I provide students with a course description with a section titled Assessment that reads:

> I am the kind of teacher who strongly believes in creating an environment where students can experience success and failure not as reward and punishment, but as information. I substitute traditional grades with informative feedback that is provided to the students while they are still learning. Report cards will be the only time students will receive a grade in my class. Report card marks will be determined based on (1) the student's projects and learning portfolio (2) my own professional judgment based on what I see and hear while observing and working with the student while they are still learning (3) self-assessments that ask the student to reflect upon their own learning and proposed grade. Assessment throughout the rest of the year will be either verbal or written feedback from me to help the student learn and improve. (Bower, 2011)

Depending on their achievement level, students respond uniquely.

The high achievers are usually a little wary of the whole idea. After all, I am removing their high. No longer can they define themselves by this metaphorical pat on the head. However, most of them are used to complying with the teacher, which all too often helped them to become high achievers in the first place. But nonetheless, they are more than a little suspicious of me.

The average achievers are mostly, if not all, appreciative for the removal of grades—perhaps only because it is something different from the monotony school has offered for so long. They thirst for something different, even if they do not really know of any alternatives.

The low achievers are so desperate for someone to cease the beatings that they nearly fall over with something that looks half like exuberance and half like exhaustion.

Stage 2: Honeymoon Sobriety

Everyone slowly but surely forgets about grades. People naturally want to learn, and kids are people, too. They are starving for an opportunity to learn for its own sake. However, for those who cannot cope with quitting cold turkey and need a grade, I do offer them the opportunity to come and speak with me in private. I then ask them what grade they think they should get. This leads to a conversation where we agree on a grade of some kind. If students care about grades, it is likely because the adults in their life are encouraging them to do so. When the adults talk less about grades, the students tend to follow their lead.

Stage 3: Withdrawal Symptoms

Everything will be going fine, and all of a sudden a few students will start to panic as if they were deep sea diving and realized they forgot their oxygen tank. They will run up to you with this ghostly look on their face and beg you to give them a grade.

It would be expected that high achievers would experience withdrawal, but you may be surprised to see how many low achievers will develop them as well. All students have been exposed to grade use for years, some as early as kindergarten, so when you wean them off of grades, they will all experience a kind of withdrawal. Regardless of their achievement level, all students, in some way or another, have come to define themselves by their teacher's judgment. So when the teacher no longer invokes their judgment, some students panic. Kohn (1999d) writes, "First, it is said that students expect to receive grades and even seem addicted to them. This is often true; personally, I've taught high school students who reacted to the absence of grades with what I can only describe as existential vertigo. (Who am I, if not a B+?)" (para. 30).

Just as a high achiever may have come to identify as an A student, the low achiever has become accustomed to identifying as an F. The point here is not that we would want more students to define themselves as A students; rather, we would want children to understand that there is more to school and learning than simply collecting A's or avoiding F's—and we do not want anyone to define themselves based on any grade. "We must constantly remind ourselves that the ultimate purpose of evaluation is to enable students to evaluate themselves. Educators may have been practicing this skill to the exclusion of learners; we need to shift part of that responsibility to students. Fostering students' ability to direct and redirect themselves must be a major goal—or what is education for?" (Costa, 1989, p. 2).

Stage 4: Sustained Sobriety

I taught my students language arts and science for 104 minutes per day, Monday through Friday, for 10 months of the year. Despite a few instances of withdrawal, my students spent more of their time focusing on learning. I never heard students ask those nagging questions like, "Is this for marks?" "Does this count?" or "Why did I only get a half-mark?" Rather than having students look at my class syllabus while doing a cost/benefit analysis to figure out if something was worth their time, they were attracted to learning for its own sake.

While it is true that I still have a curriculum and that I still have assignments that I like students to consider, I have to be flexible. Sustained sobriety from grade use means that things such as autonomy, choice, initiative, and creativity will pop up as symptoms of success. This is evident when my students ask me if they can do poster projects, blog posts, research, and science experiments. On their own initiative, they upload pictures and videos to our class forum and blog about their learning, inside and outside of school. I must be flexible, too, because I am no longer in control (and rightfully so) of their learning. We all want our students to show initiative and creativity, so do not be disappointed when they want to learn something other than what our state or province (or even you) dictates. Do not be surprised if our over-prescribed, content-bloated, externally dictated standardized curriculum gets in the way of students' learning.

During sustained sobriety, students see each other as allies to collaborate with, rather than as obstacles to be avoided or defeated in competition (Simon, 1970). They see me, the teacher, as a safe and caring ally rather than a judge-in-waiting that they must keep their distance from while showing only what I want to see. I accept them unconditionally and allow them to experience their successes and failures not as reward and punishment but as information.

In essence, I am saying that sustained sobriety from grade use can bring on an acute case of real learning, and it is awesome!

Stage 5: Relapse

At the end of grade 8, my students prepare themselves for high school. During this stage, it is inevitable for them to start thinking about how things will be different. Most of the time, they are thinking of their social lives, but many of them will ponder how their education will change.

Some will revisit Stage 3, withdrawal, as they begin to panic about high school. Some may even turn on us, their sobriety sponsors, because in their mind, they fear we have not prepared them for the rigors of high school. Some students and parents may wonder if we have actually set them up for failure. After all, we have given little to no grades, while the high school might live on them. For this, I have to share the wealth of stories I have heard from my alumni students when they come back to visit. My conversations go something like this:

"How's high school?"

"I have tons of homework. We have so much to get through in so little time."

"What is the biggest difference between last year and this year?"

"Tons of homework, and grades are important. I have to study a lot."

"What are you studying?"

"This term is mostly biology."

"What are you studying biology for?"

"To do well on the test."

Do not get me wrong, there are a lot of high school teachers who are better teachers than I am, but the system is driven to distraction—it is rotten. We have taken our eye off the ball (learning), and our children are suffering for it. So much so that even when the kids see good learning happening in class, they become suspicious (Fried, 2005) and express doubt in whether they are being "prepared" for something else.

The ultimate antidote to withdrawal late in the school year is to encourage students to reflect on their learning. I can remember sitting down with students who were worried that they were not ready for high school, and after looking through their portfolios and projects, I would ask them, "Do you think you have been learning a lot this year?" After looking at all of their evidence, they always answer yes, which leaves me with only one more question. "Can you think of a better way of preparing for high school?"

Stage 6: Mindful Reflection

Students will go to high school and experience the rigor of high-stakes tests, grades, and homework. They will play the game and jump through the hoops on their way to writing a state or provincial high-stakes exit exam. And if they come back to visit, they provide me with some interesting stories that contradict much of what they did in my grade 8 classroom. It is my experience that this step does happen, but for the most part, I may never know it, because not every student will come back and visit me. But from the anecdotal samples that I have collected from alumni students, an overwhelming majority of them experience the feeling that their learning would have been much better off if we abolished grading.

Grading without Grading

I have worked to abolish grading from my classroom as much as I possibly can. When I cannot it is because I am mandated to have a grade on the report card. It is likely that most educators face a bureaucracy that imposes some form of grading requirements. The good news is that abolishing grading or quitting teaching are not the only choices. Because the first step toward landing on the moon was taken on earth, abolishing grading entirely can happen only after a move has been made to grade less, and I am willing to bet that most teachers grade more than they are mandated. The first step to grading less is to pick an assignment or project and do not grade it. Instead, provide students with only the formative feedback they need to learn.

If you have found the courage to go a whole semester or course without any grades, you will be faced with a potentially paralyzing problem: How do you come up with a report card grade when you have no grade book? When I first faced this problem, I found myself writing to Alfie Kohn, where I outlined that I was happy to report that I had replaced everyday grading with real comments and constructive feedback, but that I was struggling with how I could convert or symbolize all the feedback as a grade, not because I wanted to but because I had to. Kohn's reply was tremendously helpful:

> My primary answer to your question is "Bring the kids in on it." This should be a decision you make with them, not for them. That goes for the general class policy (and the rationale for it) as well the specific grade given to each student. Some teachers meet with each student individually and decide together what the final grade will be. Others, who are more willing to give up control and empower students, simply let the student decide. They invariably report that students end up picking the same grade that the teacher would have given, and sometimes they even suggest a lower one. But the advantages of letting the kids decide are

> incalculable, and the process also has the salutary effect of neutralizing the destructive effects of having to give grades in the first place. (cited in Bower, 2010)

Years later, I still abide by this profound advice. Bring the kids in on it remains at the heart of my answer to anyone who asks, "How do you grade without grading?" First, even if a grade is mandated for the report card, it makes very little sense that the only way to come up with a final grade would be to take a list of other averages and average them together to get a final average (Wormeli, 2006).

The following are the three sources of information I use to establish a grade in collaboration with students:

- My students collect the evidence of their learning in their paper and electronic portfolios. The paper one is nothing fancy—just a file folder—while the electronic one might take the form of a blog, discussion forum, wiki, and/or website. It is important to note that the evidence selected for the portfolio is not just the stuff students did well; mistakes and failures are also included, not as a means to punish students, but to show growth and model that mistakes and failures are our allies in our collective pursuit of learning.
- I am a professional. I spend hours every day with my students for up to 10 months of the year. I get to know them quite well, so my professional judgment and intuitive thinking count for a lot. There is no substitute for what a teacher can see with his or her own eyes and hear with his or her own ears when observing and interacting with students while they are learning.
- I ask the students to self-assess. It is amazing how close they come to picking the same grade that I would pick. Interestingly enough, when there is disagreement, they are usually too hard on themselves, and the odd time a kid overinflates their grade, I either decide to let it go or I have a conversation with the student and make the adjustment.

What Holds Us Back?

I constantly challenge the obstacles and fears of abolishing grading, but I never disparage them—they are as real as they are abundant. Many of us have come to depend on the conveniences of grading; after all, grading can make assessment suspiciously easy. Some of the less desirable effects of grading can be the following:

- Students do not need to reflect on their own learning because the teacher will do it for them.

- Teachers do not need to worry about authentically engaging all learners because grading will garnish compliance from most and provide evidence to exclude and punish the rest.
- Parents do not need to talk with their children about their learning, because they can just ask, "Wad-ja-get?" It is too convenient to just check the grades online or wait for the next report card.
- Administrators, policymakers, and politicians do not need to engage in the messy details of what actually goes on in classrooms because they have spreadsheet-friendly data. For too many, grades have become "the only way anyone—students, teachers, or people outside the classroom—could tell how well students were doing. People stopped talking about students' abilities and interests and started talking about their scores." (Smith, 1998, p. 56)

If we convince ourselves that grading is an inevitable part of school, we might be able to live more easily with ourselves, but this translates into nothing more than willful blindness. When teachers assess with grades in a way that is said to be inevitable, we make the practice of grading inevitable and so we make the proposition true. Simon (1970) writes, "Grades must go. Their only genuine function is to serve certain administrative conveniences" (p. 397). When institutions become so large and impersonal, they become less about fulfilling their long-term goals and more about sustaining their very existence at any cost. Upon hearing "Well, that's reality—that's the system" it takes courage to stand up and say "Systems have been changed" (Simon, 1970, p. 401).

For example, it was not long ago that bullying was seen as boys being boys, as if bullying was a rite of passage (Coloroso, 2003). Today, our society has taken quite a different stance on bullying. So what happened? How did we progress from such apathy to action? I will not profess to know all the answers, but I bet it had something to do with the fact that we started to openly and actively ask provocative questions about bullying. Rather than framing the reality of bullying as something we must resign ourselves to, we started to see it as a problem to be solved. It is time we did the same for grading, and I suggest we move from asking if abolishing grading is merely realistic or practical to asking if abolishing grading is the right thing to do for children—and then do it.

Losing and Finding Our Way

For the past 100 years, there have been entire school systems that have been built on reducing student learning to numbers, and now there are education reform movements that are built on reducing the complex work of teachers to a grade (Felch, Song, & Smith, 2010; Hargreaves & Shirley, 2009). Grades were originally tools used by teachers, but today teachers are tools used by grades. This should not come as any surprise, especially if you are familiar with Marshal McLuhan, who once said, "We shape our tools and thereafter our tools shape us" (McLuhan, 1964, p. xxi). However, if there is any hope for what will likely be seen as education's data-driven dark ages, I hope it is this: It was not until the system tried to grade teachers that teachers could finally see why they should not be grading students.

The problem is not that too few students achieve A's—rather, the real problem is that "too many students have accepted that getting A's is the point of school" (Kohn 1999d, para. 8) and thus have fallen out of love with learning. When we try and reduce something that is as magnificently messy as real learning, we always conceal far more than we ever reveal. Ultimately, grading gets assessment wrong because assessment is not a spreadsheet—it is a conversation. I am a very active teacher who assesses students every day, but I threw out my grade book years ago. If we are to find our way and make learning, not grading, the primary focus of school, then we need to abandon our mania for reducing learning and people to numbers.

References

Black, P., Harrison, C., Lee, C., Marshall, B., & Wiliam, D. (2004). Working inside the black box: Assessment for learning in the classroom. *Phi Delta Kappan, 86*(1), 8–21.

Bonawitz, E., Shafto, P., Gweon, H., Goodman, N., Spelke, E., & Schulz, L. (2011, September). The double-edged sword of pedagogy: Instruction limits spontaneous exploration and discovery. *Cogniton*, 322–330.

Bower, J. (2010, July 22). Grading without grading: For the love of learning [Web log post]. Retrieved from http://www.joebower.org/2010/07/grading-without-grading.html

Bower, J. (2011, May 11). My de-grading philosophy: For the love of learning [Web log post]. Retrieved from http://www.joebower.org/2011/05/my-grading-philosophy-q.html

Brandt, R. (1995) Punished by rewards? A conversation with Alfie Kohn. *Educational Leadership*. Retrieved from http://www.alfiekohn.org/teaching/pdf/Punished%20by%20Rewards.pdf

Butler, R. (1988). Enhancing and undermining intrinsic motivation: The effects of task-involving and ego-involving evaluation on interest and performance. *British Journal of Educational Psychology, 58*, 1–14.

Coloroso, B. (2003). *The bully, the bullied, and the bystander*. New York: HarperCollins.

Costa, A. L. (1989). Re-assessing assessment. *Educational Leadership*, 46 (7), p. 2.

Coyle, D. (2009). *The talent code: Greatness isn't born. It's grown. Here's how*. New York: Bantam.

Deci, E. L. (1996). *Why we do what we do: Understanding self-motivation*. New York: Penguin Books.

Dressel, P. (1957). Facts and fancy in assigning grades. *Basic College Quarterly, 2*, 6–12.

Felch, J., Song, J., & Smith, D. (2010, August 14). Who's teaching L.A.'s kids? *Los Angeles Times*. Retrieved from http://www.latimes.com/news/local/la-me-teachers-value-20100815,0,258862,full.story

Fried, R. L. (2005). *The game of school: Why we all play it, how it hurts kids, and what it will take to change it*. San Francisco: Jossey-Bass.

Hargreaves, A., & Shirley, D. (2009). *The fourth way: The inspiring future for educational change*. Thousand Oaks, CA: Corwin.

Jones, J., & Wiliam, D. (2008). *Modern foreign languages inside the black box: Assessment for learning in the modern foreign languages classroom*. London: Granada.

Kohn, A. (1999a). *Punished by reward: The trouble with gold stars, incentive plans, A's, praise and other bribes*. Boston: Houghton Mifflin.

Kohn, A. (1999b, November). The costs of overemphasizing achievement. *School Administrator*. Retrieved from http://www.alfiekohn.org/teaching/tcooa.htm

Kohn, A. (1999c, September). Confusing harder with better. *Education Week*. Retrieved from http://www.alfiekohn.org/teaching/edweek/chwb.htm

Kohn, A. (1999d, March). From degrading to de-grading. *High School Magazine*. Retrieved from http://www.alfiekohn.org/teaching/fdtd-g.htm

Kohn, A. (1999e). *The schools our children deserve: Moving beyond traditional classrooms and "tougher standards."* Boston: Houghton Mifflin.

Lucido, R. (2012, February 27). Student learning can only be described, not measured: Living in dialogue [Web log post]. Retrieved from http://blogs.edweek.org/teachers/living-in-dialogue/2012/02/rog_lucido_we_cannot_measure_s.html

McLuhan, M. (1964). *Understanding media: The extensions of man*. New York: McGraw-Hill.

Papert, S. (1988). Seymour Papert and Paulo Freire: The future of school [Video File]. Retrieved from http://www.papert.org/articles/freire/freirePart3.html

Pulfrey, C., Buch, C., & Butera, F. (2011). Why grades engender performance-avoidance goals: The mediating role of autonomous motivation. *Journal of Educational Psychology, 103*, 683–700.

Sahlberg, P. (2011). *Finnish lessons: What can the world learn from educational change in Finland?* New York: Teachers College Press.

Simon, Sidney B. (1970, May). Grades must go. *The School Review, 78*(3), 397–402.

Smith, F. (1998). *The book of learning and forgetting*. New York: Teachers College Press.

Wormeli, R. (2006). *Fair isn't always equal: Assessing and grading in the differentiated classroom*. Portland, ME: Stenhouse.

||TWELVE

Outside the Wounding Machine

Grading and the Motive for Metaphor

John Hoben

Grading as Wounding

A couple of summers ago I taught a graduate class on biographical explorations of teaching and learning in which students are asked to create a project that reflects on how their own personal experiences have shaped their teaching lives. We talked about sticking points in our personal and professional growth, about moments of conflict and doubt, about relationships of love and caring and how all of these help to shape our identities as teachers. In the subsequent fall, I was teaching a graduate course on curriculum when I was approached by a student, Susan, a seasoned and enthusiastic teacher, who said in rather hushed tones that she wanted to speak with me after class.

I remember sitting outside in the sun listening to the telltale rustling of the leaves, when Susan began to open up about how hurt and disappointed she had felt about her grade in my previous class. I had given her an A (an 85%) for a strong submission that I felt was emotionally charged but perhaps could have been a little more critical. Recalling her final project in which she spoke of her love for her family, her faith and even the heartfelt gospel song she sang at the end, I told her that I felt that grading was subjective, that I was sorry that I had made her feel this way, and that I did value her work and her worth as a person. When I reminded her that she did receive an A I heard this seasoned teacher tell me between tears, "I can't help it. It cut me to the quick." When I asked her why she felt that way, she simply shrugged her shoulders and replied simply, "I don't know. I just had my heart set on a 90%, I guess." I told her I would read her paper again and that she should ask for a reread of the final assignment if she felt so strongly about it. Susan reassured me that she did not want to do that but told me that she felt she had to let me know how she felt because "It was eating away at me."

My encounter with Susan reminds me of the importance of the process of "associating act-and-word . . . narrative as memoir of the action that is itself a birth and an ever-renewing strangeness" (Kristeva, 2000, p. 70). Indeed, what sticks out most in my mind is my surprise at how I had unknowingly wounded her. Why such an intense emotional reaction over receiving 5% fewer marks than she felt she deserved? Despite the open atmosphere and the creative freedom that many felt the class gave them, I wonder whether she had interpreted this grade as a marker, for her of normalcy, of saying that her experiences—and perhaps she, herself—were not exceptional. Personally, I felt that grading forced me to end the student's learning experience by resorting to an act of exclusion, a casting out that makes it unequivocally clear that I am on the side of the machine. Susan was "helped, but it was in order to help [her] include the objects [s]he had made in a circuit of consumption, success measured by access"—my access to her "inner," intimate self (Kristeva, 2010, p. 253). In this sense, my encounter allowed me to see how grading functions to make us "strangers to ourselves" (Kristeva, 1991, p. 182), even as we attempt to use pedagogies of self-exploration and transformation.

For me the questions surrounding grading are incessant: Do I subtract marks for improper citation style in a paper where a young teacher talks about the death of her father with remarkable insight, wisdom, and grace? What grade do I give a teacher who has the courage to write and share her struggles with breast cancer and her fears about leaving her young daughter? Or to a young man who writes about his mother's struggles with the late stages of multiple sclerosis? Conventional grading gives no consideration to the marks these students should receive for having taught me about grace under fire, about humility and a quiet kind of perseverance instead of "sorting students like so many potatoes" (Kohn, 1994, p. 38). As a quantifying technology that presents teachers with a set of bureaucratic practices for the management of human subjects, grading is a machinery of abjection: a set of technical and administrative practices that works by "casting out" since schooling needs the threat of the wound to maintain its own internal boundaries and hierarchies. More than a simple means of disciplining students and teachers (Foucault, 1995), grading is a mode of schooling the imagination rather than allowing the imagination to radically transform schools. It does this by excluding those who do not fit prescribed models of excellence and teaching us to revile those who do not conform to dominant ways of thinking and being.

More Than Measuring Machines: Grading as a Heideggerian Technology

Despite that we rarely study the history of assessment, modern grading based on letter or numerical grades is a practice that originated in the 19th century in Harvard and Yale and was related to the desire to rank students for comparative purposes (Durm, 1993). It was also a practice that was frequently modified and can be described as "replete with trial and error" (p. 3). In the *Republic*, we might remember, Plato proposed a schooling system in which students would be led to believe that they were composed of various metals of varying degrees of value and assigned a social position suitable to their assigned worth. These metals were iron, bronze, silver, and gold and represented the workers, artisans, warrior classes, and the philosopher kings, respectively. According to this "noble lie," each class of citizens was to receive different education according to their allotted role within the state. Here, then, we have an explicit recognition of the importance of the link between social hierarchy and education since in societies characterized by limited mobility or radical inequality, individuals must be taught to accept the (in)justice of an unequal division of wealth. Plato provides us with a myth about the origins of assessment, as it were, a parable for modern grading that illustrates the power of metaphor and the dangers of taking grades as a measure of the intrinsic worth of human beings.

Yet, as the *Republic* suggests, assessment often appears to operate outside of time; it is, nonetheless, first and foremost a historical social practice and a means of creating "structured inequalities" (Singh, 1980, p. 113). Whereas preindustrial learning was centered around apprenticeship and interpersonal relations, modern learning has become a production of reducible difference centered around the grade—a process that culminated in the creation of standardized intelligence testing in the 19th century by Alfred Binet and Theodore Simon (Corbett, 2008, p. 5). Though it stands in for the knowledge of the self, a grade cannot represent the narrative and fundamentally social nature of human knowledge. Rather, as William James (1903) noted as he described the failings of American graduate education, "the institutionizing on a large scale of any natural combination of need and motive always tends to run into technicality and to develop a tyrannical Machine with unforeseen powers of exclusion and corruption" (p. 1). Writing with equal prescience more than 40 years ago, Clarence J. Karier (1969) also warned us that "American society and its educational institutions . . . have not been centrally concerned with *Humanitas*, or those values which uniquely enhance the dignity of man, but rather those which facilitate the triumph of the machine" (p. 28). As Karier suggests, the ethics of this instrumentalism and the stratification it effects is a question related to educational aims and values, one that is often avoided by treating modern assessment practices as impartial and objective.

Karier and others like him cause us to realize that standardized testing provides a technology for stratification even as it allows administrators and politicians to wash their hands of any responsibility for endemic social inequality. It does this not by sorting students on the basis of merit but by reproducing existing class inequalities, by privileging elite cultural capital as well as positivistic and rationalistic forms of knowledge. This subtle operation of power allows schools to marginalize more local and critical forms of knowledge and encourages a curriculum that ignores the need to create a democratic citizenry conscious of the need for cooperation in the pursuit of social justice. Standardized testing reminds us of the disciplinary functions of schools, insofar as they conflate credentializing with learning and reify the grade at the expense of ever-changing and evolving human potential. In this sense, it causes us to think about the nature of grading itself—namely, the ways in which its seeming inevitability and necessity overshadow a critical analysis of the ways in which it psychologically structures the types of encounters that remain open to us within today's schools.

My encounter with Susan reminds me how grading as a standardized, formal institutional practice often stands in the way of deeply imaginative and transformative educational experiences. How can educational be dialogical all the way down, if a grading *transaction* marks the predetermined end point of every educational encounter? Taking such issues into account, grading, I want to suggest, is a technology in the sense that it is a mechanism for producing, or altering a thing's being and becoming through the use of a tool or technique (Heidegger, 2004). One of the primary aims of these stratification technologies is to create the myth of machine-like inevitability in the relationship among intelligence, achievement, and social mobility. As Corbett (2008) astutely claims, "for more than a century, educators and educational thinkers have been searching for an elusive, holy grail, the edumometer . . . the set of technologies, which will once and for all, provide an objective measure of what a person has learned and what a person is capable of learning" (p. 2). In similar fashion, according to Kohn (2011), grading is "a huge, noisy, fuel-guzzling, smoke belching machine that constantly requires repairs and new parts, when what we should be doing is pulling the plug" (p. 32). Or, as he cautions us elsewhere, "Children cannot be made to acquire skills. They aren't vending machines such that we put in more homework and get out more learning" (Kohn, 2006, p. 117). These machine metaphors, like Karier's (Karier, Violas, & Spring, 1973) "myth of the mental" reflects long-standing links among grading, industrialization, and assumptions about fixed intelligence and human potential crucial to modernization and social stratification (Singh, 1980, p. 122). But more unsettling than the wasted resources that keep Kohn's nightmarish industrial machine running, that fuel Corbett's mythical edumometer, or that are at the disposal of the technocratic utopianism that Karier (1969) warns us against are the effects of grading on the lives of teachers and students. I want to build on the profound insights of the aforementioned authors by exploring grading as an abjection technology that relies on metaphorical and imaginative thinking for its power, a power that is used to create deep rooted psychological attachments, and often to wound.

Grading as a limiting technology stands in the way of the craft of teaching as an imaginative, transformative endeavor. While we tend to think of technology as the mechanized production of things, Heidegger (2004) contrasted two different types of technology: technology as techne and technology as enframing. As Heine (1990) points out, "In its initial usage, techne signified knowledge, not as the accumulation of information through observation, but the active accomplishment or manifesting realization that brings forth the illuminative power (physis) of an entity" (p. 180). Techne, involves, then, the use of a skill or art to permit the disclosure of Being, rather than transforming Being into an object or a raw power that can be dominated or possessed. In contrast, what Heidegger terms "Enframing . . . sees nature only as a reservoir of energy at man's disposal" and arises from a modern type of representational thinking that sees the world as an object to be exploited and mastered by the self (Heine, p. 180). There is a sense of technology, then, that is more akin to an art than a form of power over nature that reveals, rather than controls and disarticulates, Being (Heine, p. 181). Care, consequently, is necessary to pay attention to what technology has excluded or misinterpreted as it has disrupted our relationship with Being (Campolo, 1985, p. 435).

There is much to be wary of in Heidegger. I am not suggesting that we adopt his ontology or his reprehensible politics. I do suggest, however, that his critique of technology offers us invaluable insight into how modern instrumental rationality, evident in modern systems of standardized and criterion referenced assessment, can disrupt our relationship to the world around us and our sense of self. Applying Heidegger's critique to grading allows us to consider how grading lacks the subtlety of techne and its attentiveness to human voice and individuality; instead, it treats human beings as a standing reserve of power, a resource to be controlled and turned into an object. Rather than *bringing forth* the fundamental character or activity of a human being, grading obscures the fundamentally narrative and social nature of human existence. Within the assessment system the person becomes a quantity and, therefore, takes on a different sort of being, one that is artificial and lifeless. This occurs insofar as the subject desires the system's validation and to the extent that he or she has intimately associated his or her identity with those representations of himself or herself that the system has created.

What Heidegger shows us is that technologies make sense only in terms of encounters. Just as we consider poetry or metaphor in relation to human understanding, so too technology must have a similar focus even though most often we tend to think of the latter in terms of its end product. As Campolo (1985) puts it, "The basis of Heidegger's critique is his appraisal of technology as issuing from an aggressively self-assertive style of thought that objectifies the world in terms of its potential usefulness to the human subject" (p. 435). Techne's power comes from its ability to reveal truth and its capacity to alter the relationship between humans and the world that they inhabit. Techne, like the creative arts, permits the disclosure of Being; it enables rather than distorts the process of allowing things to become what they are meant, are intended, or intend to be. Like poetry, techne opens up a range of possibility by revealing something about the relationship between humankind and the world. According to Campolo (1985):

> This reading of techne suggests a certain relation to poiesis. "In this original sense of the word," Harold Alderman points out, "a technician would be a kind of poet who succeeds, in an originative uncovering of beings, that is to say in a disclosure of a particular world. . . . [T]he fundamental difference between the original techne of poiesis and the modern techne of technology is that the first is responsive and contemplative whereas the second is domineering and challenging" (35). The problem with technology, then, seems to lie in the suppression of techne's original meaning. Modern technology assumes "responsibility" for beings only insofar as it takes credit for inevitably violent human manipulation and manufacturing. But this aggressive style excludes the original sense of techne that points toward a care for Being that permits the poetic uncovering and lighting up of beings. From this perspective, technology must, paradoxically, be criticized for not being "technical" at all. Where genuine techne is approached, building is already dwelling and the technician is also a poet. (p. 436)

Techne, then, deals with "unconcealment" or heightening our attentiveness to the countless ways in which we experience, and are able to access, the world. Taking the transactional educational process created by grading as a substitute for this process of engagement then is to fundamentally undermine one's ability to pursue knowledge. Like enframing, grading renders human consciousness into an abstract quantity, and thereby "allow[s] each person to emerge as a 'case', an analyzable object" (McClam & Sevier, 2010, p. 1462). In this way, grading forces us to identify our interests with this technology of concealment, even as it threatens to wound us psychologically by casting us out should we fail to comply with the demands of the assessment machine. Critical pedagogy causes us to question taken-for-granted assumptions about assessment as we strive to create a poiesis of teaching that is grounded in an acute institutional literacy. Despite ongoing efforts to deskill the profession, as critical teachers we are accountable first and foremost to our craft and our professional duty to develop the inherent potential that exists within every classroom.

Abjection and the "Self on Trial"

Schools, like capitalism itself, need people who fail. When, as teachers, we take grading for granted we ignore the disenchantment and pain created by grading as a ubiquitous institutional practice. Our inability or unwillingness to recognize the operation of power at the grade level reveals our complicity in the institutional structures that reproduce social inequality through rituals of power and passivity. For such reasons, de-testing and de-grading schools is more than just a means of providing flexibility in assessment for teachers—it is about broadening the scope of the fight for social justice to include all aspects of schooling.

Reimaging grading is the first step toward creating educational spaces that are filled with hope and wonder rather than the intense self-consciousness that is the unfortunate legacy of a politics of accountability. Much as Illich (1971) called for the deschooling of society more than 40 years ago, the call to de-test and de-grade schools would enable us to confront the cold calculating reasoning that has held wonder hostage and prevented us from experimenting with new forms of assessment that respect and enhance human potential. Seen within this broader institutional context, Heidegger provides us with a theory of becoming in schools that are about shaping students' growth and individual transformation. Yet, Heidegger can show us how the notion of encounter is central to finding authentic ways of being and becoming through a heightened sense of imagination and care, and this still requires us to consider the types of harm that arise when encounter is foreclosed. That is, if grading forces us to be in certain ways, what are the personal psychology foregrounding experiences of estrangement?

Grading, particularly in its standardized forms, has become a disciplinary instrument rather than a tool that enhances and furthers the educational experience. In many ways the massification of educational practice causes us to forget the negative formative effects of grading rooted in a deficit-based model of public education. In this vein, I want to suggest that the idea of abjection developed by Julia Kristeva, the French-Bulgarian psychoanalyst and cultural critic, shows us the psychological mechanics through which the artificial modes of being promoted by grading harm students. *By sorting students and by expelling those who are somehow deficient, grading operates as a social and psychological instrument of abjection.* Indeed, for Kristeva, abjection is one of the key processes by which the self maintains its own internal boundaries. The original abject object is the mother that the infant must reject if he or she is to become an individual, even though at some level the infant is still dependent on the mother to meet his or her needs. As a result, this process of rejection is always accompanied by an unrecognized desire for the object of need, a recognition that is at the limits of consciousness and is always threatening to resurface and to undermine the ego's integrity. The abject in this sense represents the pre-linguistic state in which the subject was not fully differentiated and thus represents the threat of a dissolution of self.

Since it is a fundamental psychological process we cannot do away with abjection, although we can be sensitive to its social and personal effects. Kristeva's "self on trial" is continually in the process of contestation and revision, yet abjection lends us the illusion that it is stable and permanent. By associating that which is outside of the self with feelings of horror and disgust, abjection is a psychological mechanism by which societies and individuals expel that which is *other* a process that at once establishes powerful norms and institutes a process of self formation. Abjection allows the subject to live in the comforting illusion that it is whole and stable (Danforth, 1997). In this way the experience of abjection can also represent the path to a deeper self understanding—emphasizing the importance of being attentive to what is *not there* since, according to Kristeva (1982), "all abjection is in fact recognition of the *want* on which any being, meaning, language, or desire is founded" (p. 5). Schools need failures, then, because, in a sense, they are its proverbial heads on sharpened sticks—or, at the least, the ignominious and lifeless scarecrows that mark the rigid boundaries of modern schooling.

Can we practice critical pedagogy while holding onto the cold conventional calculus of instrumental assessment? The need to de-test and de-grade schools is, in part, intimately connected with a recognition of the deeply cultural nature of human learning. Kristeva, we should note, also insists that the standards of value through which abject objects are identified and defined are often social (Danforth, 1997, p. 6). Here, then, we have a social system premised upon a logic of hierarchy where what Bauman (2004) calls the "horrors of exclusion" encourage unquestioning compliance (p. 128). That abjection has a social dimension means that it plays an important role in bullying, group identity, discrimination, and other forms of social behavior that rely on the creation of starkly contrasting forms of difference to create group solidarity. In many ways, the marginalized in this case "are constructed as psychological foils submerged along a gradation of moral value" (Danforth, 1997, p. 6). These boundaries are sometimes overtly enforced through what Danforth calls a status degradation ceremony, a social rite in which an individual is singled out as an example of the abject and humiliated by the *normal* group plays not only a powerful role in bullying and the creation of marginal groups such as the poor, but also in many forms of prejudicial, racist, sexist, and homophobic thinking. In such a system "the new Big Brother's concern is exclusion—spotting the people who 'do not fit' into the place they are in, banishing them from that place and deporting them 'where they belong' or better still never allowing them to come anywhere near in the first place" (Bauman, 2004, p. 132). Not surprisingly, the anxiety students feel at the threat of abjection animates the psychic life of the classroom, rather than allowing teacher and learner to open themselves to the transformative power of love, hope, and the human imagination.

But as critical teachers who have to navigate rigid institutional hierarchies, what options remain open to us? Opposed to abjection, Kristeva suggests the need for an awareness of the processes through which identities are created and the ways in which selves are shaped through an interaction between language, being, and desire. Despite the pretense of modern positivism and empirical science, Kristeva maintains that the psychic dimension of human life or what she terms the semiotic aspect of language can never be excluded from the symbolic meaning of words. In the modern world Kristeva sees the need for something she terms "intimate revolt," a form of rebellion that is aware of the subtlety of power and the vulnerability of others. For Kristeva, revolt is an act, perhaps an art, through which the individual remains aware of the fact that power is always exercised in spaces in which the lines between agency and oppression are not so easily drawn. Within such a conception of society, the importance of the liminal spaces and meanings grows since abjection refuses ambiguity, it requires meaning to be fixed: "What does not respect borders, positions, rules. The in-between, the ambiguous, the composite" (Kristeva, 1982, p. 4). Navigating these proximal spaces "awakens a catastrophic anxiety that in turn leads to defensive reactions of rejection, indifference, or arrogance, when not the will to eradicate" (Kristeva, 2010, p. 258) and the construction of social notions of "normal" and the "deficient." The abject really is replete with danger, then, because it can become an incubation point for what Hannah Ardent called the "behavioral self": "one whose experience of life is habitual, who responds to the world in conformist terms, who is not acting, or making, but as best laboring" (Dumm, 2008, p. 51). Instead, critical teachers recognize the necessity of taking the opportunities present within the little moments of the teaching life, to practice intimate-critical forms of pedagogy that treat each student differently and gives voice to the repressed longing for a world that is at once more caring and intensely human (Kelly, 1997).

Identification and the Wound: The Motive for Metaphor

Finding these moments of resistance on the margins is not always easy. Indeed, those who wound others so deeply need good alibis. Recognizing the need to de-test and de-grade schools arises, in large part, from the realization that the accountability movement is based on the misconception that imagination, metaphor, and one's personal subjectivity can be divorced from assessment—the pervasive yet erroneous belief that judgments about competence are exclusively technical and objective rather than creative, experiential, and personal. Instead, I want to suggest, even the most seemingly technical and objective forms of assessment are intimately related to positionality and the human imagination.

Despite what standardized assessment practices teaches us, there is no such thing as a common or an average human experience. Both Kristeva and Heidegger remind us that human existence is relational and disclosed through language that helps us to gain some sense of how we find our place in the world. Heidegger's enframing, like Kristeva's concept of abjection, offers a powerful way of understanding the human desire to dominate, to treat others as mere waste or stock manifested in modern systems of grading. Both thinkers challenge us to reimagine the world we see and feel when we are most at ease with ourselves. By confronting us with a type of uncanny strangeness, they show us the undying need for intimacy, imagination, and care in everyday human experience. Language, they suggest, moves us far beyond the paths set out by human intention—to fail to recognize this is to find ourselves continually mesmerized by the steady, slow catastrophe of a self sated by its own mute ability to possess what it claims in the waning half light of the conscious mind.

Why is grading so rarely challenged? Symbols and metaphors condition, at a fundamental level, how we see ourselves. Yet, what we want and intend are not always clear, nor the same, though they are always closely connected to who we are and who we can become. In a sense, grades present us with doubles, model ideals of how we are, and are not, supposed to behave—"grades speak for and on behalf of people" (Roth & McGinn, 1998, in McClam & Sevier, 2010, p. 1461). Assessment is a type of thinking that attempts to situate itself outside of imagination by limiting the scope of metaphor to one of simple numeric or alpha-numeric equivalency. Unlike poetry and narrative, which is "living speech" (Kristeva, 2000, p. 57), grading is a system of strict equivalences; there is no room for nuance or imagination; it is strict, orderly, reductive, and mechanical. Grading at its heart is metaphorical, since it derives its power from its ability to convey relationships of containment, linearity, and hierarchy by invoking images such as number lines, discrete sets, the practice of sorting or assigning differing values to objects (i.e., products or goods), the simple processional letter board ordering of the common Roman alphabet. For these reasons, assessment reminds me of Wallace Stevens's (1990) poem "The Motive for Metaphor" as we consider the cost of the orthodox view that all educational encounters begin and end with the grade: "The wind moves like a cripple among leaves / And repeats words without meaning" (ll. 3–4, p. 288).

Frye (1993) talks about this poem in terms of correspondence: the desire for some deep communion between self and world, mediated by language. For me Stevens's poem deals with the language of rejection and death, a type of melancholia that is suggestive of the depth and subtlety of human feeling and the loss that inevitably accompanies existence. Yet, despite this, the answer to this incessant desire for answers is not instrumental logic that turns people and beings into mere objects. If there is any answer, Stevens suggests, it lies in the unfolding of language, and the ability of metaphor to insinuate other forms of being and knowing. There is both gravity and inertia to *being* that is responsive to language, in ways that are both poetic and necessary—perhaps this primordial melancholia is a part of our own being in ways that can be sensed but never quite fully articulated. This sort of resistance to capture can never be fully overcome by what Stevens calls "The vital, arrogant, fatal, dominant X," itself a metaphor, perhaps, for the totalitarian mind that seeks to impose its will, to possess, but can only displace or destroy the lost object of its desperate desire.

Both dialogical education and assessment are fundamentally creatures of language, of metaphor and imagination that frame the broader conditions for all learning experiences. For teachers and student alike, the motive for abandoning metaphor is the fear of encounter and the ensuing desire to take assessment's objectified subject as a replacement for a caring, feeling, narrative self. To kill our common humanity, authoritarianism must first kill our capacity for creative, life-giving language; it must steal our wonder (Doyle & Hoben, 2011). To do so requires social practices that restrict and constrain imaginative language, that mislead us into thinking that language can be neatly disassociated from the unending word-and-world-play that collectively constitute the ongoing act of being human. Yet, while instrumental logic can constrain creative language, it can never replace it. Indeed, abjection itself requires metaphor because it relies on associative thinking to tell us what to revile and reject; like Being and technology itself its primary nature and function is relational. As Frye (1993) emphasizes in *The Educated Imagination*, "As for metaphor, where you're really saying 'this is that', you're turning your back on logic and reason completely, because logically two things can never be the same thing and still remain two things" (p. 16). Metaphor redefines boundaries, "transform[s] thought into phenomenon, reconciling it with perception and common sense" (Arendt, cited in Kristeva, 2000, p. 69).

Although grades are static and quantifiable, learning is fundamentally episodic, experiential, and narrative in nature. Rather than quantifying human beings, critical teachers need to examine (or assess) the degree of fit that exists between human experience and schoolings' overarching structural metaphors. We need to question the system's deep abiding faith in Stevens's "vital, arrogant, fatal, dominant X" as we learn to see the tragic human costs of an accountability movement radically divorced from both the everyday realities faced by teachers' and students' curious and imaginative minds. This awareness is the backdrop for an expanded sense of care, one that opens the way for a new age of intimate revolt in schooling and that is capable of distinguishing between learning interactions that genuinely transform students and those that wound through disciplinary tactics. If we remember Heidegger: "the critique of technology culminates in a turn to the metaphors and imagery of guarding, keeping, questioning, waiting, and caring" (Campolo, 1985, p. 400). To resist insidious assessment technologies that make certain types of being possible and others impossible requires both an awareness of the types of wounds grading causes as well as the subtlety with which it co-opts imaginative thinking. Undeniably, then, there is a paradox to grading because while it purports to exclude the imaginative and the subjective, it is dependent upon the individual's ability to imagine that one thing is another: one among a selected and limited number of fixed possibilities. Grading makes feeling and the imagination the abject, since these are not part of the raw power that grading wants to create from society's vast reserves of human capital, causing us to neglect the more fundamental question of *who we are*.

As Susan discovered, grading technology allows schools to exist as abjection machines that create a stratified, competitive society in which we learn that the price of not towing the line is humiliation, psychological insecurity, and exclusion. Her story is an indication of how grading can act as a status degradation ceremony of the sort described by Danforth, one that allows us doing the humiliating to feel as though we can live with clean hands because there is a wounding machine that does the dirty work for us. There is, as we have seen, always a story behind the grade, often one that hints at the need to trust in the power of caring and the imagination put into service in the name of friendship and love. Instead of quantities and abstractions, metaphors allow us to see and understand the world-in-language, as we evoke the experiences, feelings, and stories of those we cherish and love. In this way, one needs metaphors that are generative and possibility-raising rather than a means of limiting human understanding. Those possibilities can be a source of strength or, if they are denied and repressed, a source of frustration and weakness. Posing as a final, definitive representation, the grade, to retain its power over human affairs, must hide the narrative, situated, and local nature of human knowledge. Why bother, then, with measures and dead numbers rather than seeking to discover how we can make ourselves become better than the other? Why not make learning about growth and acceptance and the struggle for shared understanding? Perhaps, I want to suggest, we have been too easy on ourselves as critical educators, since we have failed to direct our collective attention to our own complicity in the ongoing travesty of conventional grading and testing—a glaring failure that continues to plague our students and schools today.

Paper Planes: The Lost Poetry of Education

One of my favorite movies, *The Paper Chase* (1973), portrays the intensely competitive world of Harvard Law School and the personal trepidation felt by those who are unable to meet the mark. Among the most memorable characters is John Houseman's portrayal of Professor Kingsfield, a first-year contract law professor who is renowned for his cruelty and his ability to *dress down* unsuspecting first-year students. The conception of education we find here is one rife with condescension and disdain. In Kingsfield's words, "You teach yourselves the law, but I train your minds. You come in here with a skull full of mush; you leave thinking like a lawyer." James T. Hart, played by Timothy Bottoms, is a student in Kingsfield's class who attempts to retain his dignity, to earn good grades and to win both Kingsfield's approval and the love of his young daughter, Susan. Despite much personal angst, Hart does eventually shine as a student and appears to be on his way to law school success. However, in an unsuspecting twist, the movie closes with Hart on the beach with Kingfield's daughter folding an envelope of his first-year grades into a paper plane and tossing it into the ocean. While Hart does seem to recognize the immense personal cost of the "paper chase" and the sterility of a life filled only with academic learning, throwing the unopened grades into the ocean is an act of revolt that reminds us that "it is in action, as capacity for beginning, that the human condition of individuation is actualized" (Kristeva, 2000, p. 58).

Grading, I suggest, is something like a neat trick that power plays on us to rob us of our agency. While its logic is instrumental and mechanical, it is fashioned around an old myth: namely, that obedience and devotion to the system will bring self-fulfillment. Machine logic creates a cultural amnesia whereby we forget that "education is just shorthand for an enormously complex set of different strategies, and not a single button on the machine to be pushed or not" (Banerjee, 2007, p. 14). As Stevens so eloquently reminds us, there is no transcendent reality to which consciousness belongs that is divorced from everyday existence and certainly none that can be encapsulated by a single letter or number. Rather, the search for transformational education is the search for "a project in which my fear of . . . defect and death . . . is transformed into attention, patience, and solidarity capable of refining my being in the world" (Kristeva, 2010, p. 266).

Where will our totemic fetish for the lifeless machine lead us? It is, I contend, crucial to recognize that the desire to abdicate one's agency to the machine is at some level indistinguishable from the desire for death. However, more often the machine fetish takes on the form of a desire for a simple totalitarian logic obsessed with a simple conceptual tidiness, of putting things in their proper place. As Banerjee (2007) points out, "the machine question" is the misguided search for the universal law that will bring certainty to the search for causes and effects in the social sciences (p. 14) even as neoliberalism remains fixated on "the idea of an economic machine, self-perpetuating and beyond human volition" (p. 12). As part of this abject logic, grading casts the nonconformist and the original thinker out, and prevents us from seeing that this is neither the result of a lack of merit or virtue. It makes us turn away from the lost poetry of education, a poetry that inspires us to dare to dream of a tomorrow where learning is not accompanied by the threat of wounding or casting out.

Grading is designed to create students who fail. Students who fail validate the knowledge of the masters behind the curtain who pull the invisible levers of the relentless unthinking machine that makes the world go round, a machinery that we are told is necessary and just but which really runs on human fears and ruined dreams. In this sense, failing grading requires recognition of the deep wounding this machinery produces: intimately connected with the idea that masses of people are not smart enough or good enough to run their own lives.

It is a machine that diagnoses the disease that it is engineered to create, a technology that creates failures and allows us to deny that these failures are our own failures. The construct of the *good* or exceptional student hides our complicity in creating the shadow side of this ideal—the failure that stems from our ignorance of ourselves and our own fallibility. Rather than seeing the arbitrariness of the dividing line between success and failure within modernism and its institutions, grading shows us how "the defense against the abject is bolstered by symbolic investment in the ego-ideal, which may embody stylized or exaggerated ideas" (Linstead, 1997, p. 1116).

Rather than seeing assessment as the end of every educational encounter, perhaps we will come to see that behind every grade there is an unspoken story and, often, a hidden wound. Rather than pretending that all of our human weakness can be cancelled out by the act of knowing, by admitting vulnerability we become stronger, more caring, and open to seeing the poetic in the everyday. This is an aspect of the human condition that forces the individual to draw on experience, imagination, logic, and love—indeed the whole range of our encounters with the other—and that hints at the necessity of confronting the existential—or, some might say, spiritual—questions that define us. Those who fail are often the ones who can teach us how to really make the grade: that crucial lesson about the importance of learning the need to help others to excel instead of being a means of casting others out; to borrow Stevens's phrase, the true measure of "things that would never be quite expressed." Grades, we should remember, are only floating markers on an ocean that is vaster and more powerful than we can ever imagine. To ignore the cost of grades and the wounds they impose is to resign oneself to being marooned in ignorance, no new horizons ever coming closer—only a harsh future looming like the incontrovertible stare of a blank and pitiless sun. It is time, then, to put aside the language of machines and learning simply how to become more fully and genuinely human—time to find grace and wisdom in our heartfelt stumbling, in a world obsessed with limitations and the need for wounding.

References

Banerjee, A. V. (2007). Inside the machine: Toward a new development economics. *Boston Review, 32*(2), 12–18.

Bauman, Z. (2004). *Wasted lives: Modernity and its outcasts*. Cambridge: Polity Press.

Campolo, L. D. (1985). Derrida and Heidegger: The critique of technology and the call to care. *Journal of the American Academy of Religion, 53*(3), 431–438.

Corbett, M. (2008). The edumometer: The commodification of learning from Galton to the PISA. *Journal for Critical Education Policy Studies, 6*(1). Retrieved from: http://www.jceps.com/index.php?pageID=article&articleID=125.

Danforth, S. (1997). Autobiography as critical pedagogy. *Teaching Education, 9*(1), 3–14.

Doyle, C., & Hoben, J. (2011). No room for wonder. In J. Kincheloe & R. Hewitt (Eds.), *Whatever happened to soul? A manifesto of revival* (pp. 115–127). New York: Peter Lang.

Dumm, T. (2008). *Loneliness as a way of life*. Cambridge: Harvard University Press.

Durm, M. W. (1993). An A is not an A is not an A: A history of grading. *The Educational Forum, 57*, 1–4.
Foucault, M. (1995). *Discipline and punish: The birth of the prison*. New York: Vintage.
Frye, N. (1993). *The educated imagination*. Toronto: House of Anansi Press.
Heidegger, M. (2004). *The question concerning technology*. New York: Harper Torchbooks.
Heine, S. (1990). Philosophy for an "Age of Death": The critique of science and technology in Heidegger and Nishitani. *Philosophy East and West, 40*(2), 175–193.
Illich, I. (1971). *Deschooling society*. New York: Harper.
James, W. (1903). The Ph.D. octopus. *Harvard Monthly, 36*, 1.
Karier, C. J. (1969). "Humanitas" and the triumph of the machine. *Journal of Aesthetic Education, 3*(2), 11–28.
Karier, C. J., Violas, P. C., & Spring, J. H. (1973). *Roots of crisis: American education in the twentieth century*. Chicago: Rand McNally.
Kelly, U. A. M. (1997). *Schooling desire: Literacy, cultural politics, and pedagogy*. New York: Routledge.
Kohn, A. (1993). *Punished by rewards: The trouble with gold stars, incentive plans, A's, praise and other bribes*. New York: Houghton-Mifflin.
Kohn, A. (1994). Grading: The issue is not how but why. *Educational Leadership*, 52(2), 38–41.
Kohn, A. (2002). The dangerous myth of grade inflation. *The Chronicle of Higher Education, 49*(11), 7.
Kohn, A. (2006). *The homework myth: Why our kids get too much of a bad thing*. Cambridge: Da Capo Press.
Kohn, A. (2011). The case against grades. *Educational Leadership, 69*(3), 28–33.
Kristeva, J. (1982). *Powers of horror: An essay on abjection*. New York: Columbia University Press.
Kristeva, J. (1991). *Strangers to ourselves*. New York: Columbia University Press.
Kristeva, J. (2000). *Crisis of the European subject*. New York: Other Press.
Kristeva, J. (2002). *Intimate revolt*. New York: Columbia University Press.
Kristeva, J. (2010). Liberty, equality, fraternity, and . . . vulnerability. *Women's Studies Quarterly, 38*(2), 251–269.
Linstead, S. (1997). Abjection and organization. *Human Relations, 50*(9), 1115–1145.
McClam, S., & Sevier, B. (2010). Troubles with grades, grading, and change. *Teaching and Teacher Education, 26*(7), 1460–1470.
Singh, A. (1980). Equality of educational opportunity. *Guru Nanak Journal of Sociology, 1*(1–2), 109–138.
Stevens, W. (1990). *The collected poems of Wallace Stevens*. New York: Vintage Books.

|| THIRTEEN

No Testing Week

Focusing on Creativity in the Classroom

Peter DeWitt

"We are raising a stressed out generation of students who are over-tested and over-analyzed."

When I entered college, a friend's parents, who were both teachers, tried to dissuade me from entering the field of education, which I found very sad. They were both excellent teachers, but they said the profession was changing, and not for the better. I politely smiled and listened to their concerns, but I continued down the same path despite their warnings. After working in an after-school program, I knew that I wanted to be an educator. I never forgot the disappointment I felt when those two retired teachers tried to talk me out of entering the profession that they spent so much time in. After 17 years in education, first as a teacher and then a principal, I understand why they felt the way they did so long ago. However, I still maintain hope that things will get better and strongly believe it is my job as the school leader to help teachers find that love again.

I have come to a crossroads in my career. According to the movie *Under the Tuscan Sun*, that sounds very Oprah of me. When I began teaching I remember more seasoned teachers stating that if you stay in education long enough, you will see the pendulum swing from one side to the other. It is my hope that the pendulum has swung to one very dysfunctional side long enough and will make its way to a side that is based in common sense and sound educational practices before many of us end our careers.

It seems as though policymakers in education want educators to pay attention to research, data, and accountability, but they feel that they do not have to play by the same rules. Apparently research, data, and accountability matter only when they tell policymakers what they want to hear. Unfortunately, the direction in which they have been leading education is not good for kids; it is bordering on educational malpractice. Just like the present economic issues in the United States, education will continue to benefit only the top percentage of kids who can afford it.

Rothstein (2004) states, "On average, professional parents spoke over 2,000 words per hour to their children, working class parents spoke about 1,300, and welfare mothers spoke about 600. So by age 3, children of professionals had vocabularies that were nearly 50% greater than those of working-class children and twice as large as those of welfare children" (p. 28). Those students who have the luxury of being exposed to rich experiences are the ones who enter school far ahead of their peers. Their experiences allow them to understand material that other students cannot fathom. High-stakes testing was once considered a way to level the playing field, but it has done nothing more than make some students feel like failures because they cannot get a 3 on a test. In addition, in many states those same high-stakes tests are used for as much as 50% of a teacher's evaluation.

Data-Driven Decision Making

> *Education, of course, is overloaded with programs and data. The growth of digital power has aided and abetted the spread of accountability-driven data—adequate yearly progress, test results, for every child in every grade, common core standards, formative and summative assessments galore. (Sharratt & Fullan, 2012, p. 2)*

Educators are under increasingly intense pressure to maintain data on their students. They formatively assess to make sure that their students are keeping pace with peers and then give summative assessments to make sure they made progress. Principals used to have high-stakes testing data only to show how their school was doing in a moment in time, but now it is provided on any given day and at any given time. Unfortunately, it is also being used to show the effectiveness of teachers and administrators.

Besides *data*, one of the words that educators hear a great deal is *accountability*. Educators are held accountable for the progress of their students. In the past 10 years since the passage of the No Child Left Behind (NCLB) Act, schooling has changed from a place of creativity to an institution that needs to constantly prove the students are learning. As much as data used to center around high-stakes testing, schools are keeping other forms of data to counterbalance the effects of high-stakes testing because those are the data that schools mostly hear about in the news or from education departments at the state and federal levels.

High-stakes testing will be the ruination of public education as we know it. At this point, many states have spent years forcing high-stakes testing on schools and few of these schools have shown improvement. The unfunded mandates and paperwork that have come with NCLB and Race to the Top have forced schools to focus on red tape rather than students. In addition, they are bringing down the morale of teachers around North America. Metlife and Harris Interactive partnered on the *Metlife Survey of the American Teacher* (Markow & Andrea, 2012) and found that teacher job satisfaction is at an all-time low.

Testing is something that educators do not have full control of when the timer begins. We know that there are many outside influences that affect the outcomes of the tests. Kids have bad days, come from households that do not support education, and many students suffer from test-taking anxiety, which has gotten worse over the past decade (Sadker & Zittleman, 2004). This negatively affects their performance.

Schools can learn from the problems with high-stakes testing by lowering their focus on summative assessments and spending more time focusing on formative assessments. Formative assessments through the use of observation, discussions, exit polling, and learning and response logs can give educators as much information about student learning as summative assessments. In addition, formative assessments are often less stressful for students. When done correctly, the use of assessment can be beneficial to educators and administrators. Using high-stakes testing to tell whether a teacher is successful or not just doesn't work. What once may have started as a way to monitor schools has quickly become a political tool to say that schools are failing.

High-stakes testing has changed since the first days of NCLB. There was a time when state education departments would ask for feedback from school districts, and they would work together to create tests that had a variety of testing questions for a variety of learners. With increased accountability, schools no longer have as much input from state education departments. It has become so combative that schools have to send back any unused testing booklets because state education departments do not want schools to be able to use them for test prep. Schools are accountable for every test booklet they request.

Educators are now teaching during a different time in education, where what is tested is what truly matters. Classrooms are less creative, teachers are more stressed, and students are walking into classroom environments where they are expected to perform at any moment. If they do not perform well, that may have dire consequences for the teachers and administrators who will be evaluated based on student performance.

In addition to all of the testing stress there have been a few examples where educators have had their names and Value-Added Assessment scores printed in major newspapers. Teachers from the Los Angeles Unified School District and New York City educators have both found their scores printed for the public to see, even though there were huge margins of error, creating frenzy in both cities as well as across the country.

A Week without Testing

During the fall of 2011, I took some time to craft an email in a Word document. I needed to make sure that I chose my words correctly because I wanted my message to the staff to be clear. I, along with thousands of other teachers, principals, and parents, am becoming increasingly concerned about the direction that education is heading in the United States. There is a focus on testing that we have never seen before, and it is ruining creativity in the classroom.

As a school principal, sending the whole staff a message is something that I take seriously because once your words are out there, they can be interpreted in numerous ways. I did not want to write anything that staff would find offensive, but I did need them to understand the urgency of the situation. The reason for the email was to communicate something that I feel strongly about. It had to do with the overuse of testing in the United States and my desire that we focus more on creativity in our school.

Once a month, I meet with my Principal's Advisory Council (PAC). I have two co-chairs who are teachers within the building. They are open and honest, even when they are saying things I may not want to hear. PAC is not about venting about building issues, but about meeting to discuss how we can improve our building environment. I wanted to approach PAC about having one week that is test free. After all, do we really need to test students every week? I decided to send the staff an email prior to PAC because I wanted them to understand where I was coming from, and hopefully be on board with my message.

As a principal and educator, I am concerned that all we ever hear about is testing. Our scores are available online to anyone who wants to see them. However, our school environment is not available for everyone. The happiness and engagement levels of our students are not available either, so in the end, clicking on a link that says *See How Your Kids Are Doing* really means *See how your kids are doing in one particular area that took place over a three-day period.*

I am fortunate because I work with great staff and awesome kids, but I worry that we are only measured by a test and not by our creativity. I want our kids to live and breathe creativity all the time, but I need to begin with one week. Just one week to open up new doors for them. One week without test anxiety. Perhaps we would even outlaw the word *test*. I finished the draft of the email, closed my eyes, and pressed send.

Thankfully, I received some quick responses from staff who applauded the decision. They were not sure what they would do during the week but they were fine with making sure they did not test. Other teachers were not as open. At PAC we discussed the list of testing that would be outlawed for the week. There would not be any progress monitoring, science tests, or spelling tests.

After much discussion at PAC, the staff went their separate ways and kept discussing the week. A day later I posted a blog about No Testing Week on Finding Common Ground, which is my blog for *Education Week*. Within a day the blog was sent out on Twitter more than 100 times. I heard from educators around the world. Suddenly, No Testing Week took on a life of its own, and I was excited to see where it would go.

However, I was also a bit nervous. As a school principal I have a duty to make sure the rules set by the New York State Education Department are followed. Speaking out against testing was something that only a few people had done, and most of them were not practicing school principals. Larger then my duty to follow the rules set by the state education department is my obligation to students and staff. Testing is harmful and I wanted the world to know what I thought.

No Testing Week

During the week of November 28 through December 2, 2011, our school participated in No Testing Week. Instead, we focused on doing projects and other creative activities. Our school usually participates in two Scholastic Book Fairs and the week that brought together November and December is one of the weeks Scholastic was at our school. Our students bought books all week long. They were surrounded by books. They had extra time to get lost in the wonder of their favorite book. On Friday evening, December 2, local children's author Matt McElligott gave a presentation and read to children and families.

The reasons for doing this were plentiful. In the United States we are too focused on testing and I strongly believe the only way to bring back creativity is for principals to give teachers permission to spend time without worrying about data. Good data that inform instruction will always be important, but I do not believe we always collect good data. I also believe we are raising a stressed-out generation of students who are over-tested and overanalyzed.

Over time, momentum caught on and teachers became ecstatic. They were on board with spending a week without data. They were happy to be given free rein to focus on projects and other creative activities that are highly effective in building student engagement. I could feel the tension in the building slip away. The following are *some* of the activities we did:

- Reader's theater involving one of Matt's books. Teachers and students performed for the whole school.
- Our librarian built a pit (swimming pool) in her library, which is the setting of Matt's book *Uncle Frank's Pit*. Students filled the pit with different objects by the end of the week.
- All students and staff wrote about their favorite book and hung their pieces around the school.

- Matt has many books about pirates and our third-grade teachers did a week's worth of pirate activities and lessons.
- All three fourth-grade classes built a long house with their students (Social Studies curriculum). Every class was responsible for building a section (see Picture 13.1).

Picture 13.1. Student-built section of long house.

- I digitally recorded myself reading books, and our librarian set it to a PowerPoint. The PowerPoint was made into a movie with graphics. During the day teachers went to the shared folder, brought up the PowerPoint, and listened to me read the story.
- We read *Bean Thirteen* and made mosaics.
- Teachers used Matt's book *The Lion's Share* to teach students measurement and fractions, which involved some baking.
- All teachers were involved in project-based learning projects with their students.
- Our school was transformed into a creative environment for learning, which will spur more ideas.

I understand that we could do these activities regardless of testing. However, the building environment changes when there is testing involved. People are less patient and more stressed. We know testing is our reality, but for one week it was the furthest thing from our minds. We plan on finding other weeks where we can go without testing. After all, it's an elementary school, and all elementary schools should be places that spark the imagination and not put it out.

National Story

I received a phone call from a senior writer at *NEA Today*. The writer had contacted a few of my staff members to make sure that I was "walking the talk," which is always a great idea. Although I blogged about participating in a no-testing week, my readers were not sure if I was really going to do it. Perhaps I was just making a point or trying to send a symbolic message. Some of the teachers assured the writer that we were planning the week. Previous to when we were hit with NCLB, elementary schools were (and still should be) a place that is creative and fun. We had fewer parameters, and we were less concerned about a daily focus on textbooks. We had a better balance of education and fun.

The point of No Testing Week was to go back to a time before NCLB and all of our mandates. In addition to the benefits of a less stressful week and focusing on creativity, we have seen an increase in the sense of community within our school, although we have always been a close staff, and we have had open and honest discussions about testing, NCLB, and how we can move forward in a time with so many mandates.

Something amazing happened after I posted the blog in October. I heard from educators from across the country and around the world. There were many educators like us who wanted something more for their students. They wanted a week where they could focus on student-centered topics and expose students to different types of learning. It is my hope that many schools will create their own No Testing Week and allow their students to have a week that is free of the stress from testing.

Testing vs. Assessment

After the blog post about No Testing Week, I heard from many educators, some of whom wanted me to know the difference between testing and assessment. It was a difference that we already knew in school, but I understood where the educators were coming from; they worried that I was suggesting that assessment was bad.

Testing and assessment are very different. We can assess students' understanding of concepts. Assessment is a powerful tool that all educators need to understand. Testing, however, is something different. It may have started as a powerful tool to help guide instruction, but it has turned into a corporate-driven, top-down decision to prove that schools are failing, regardless of whether they really are or not.

In a time when all we seem to hear about in education is high-stakes testing and that schools are failing, I wanted to take the opportunity to look at our students differently. It gave us the occasion to watch all of our students shine. No Testing Week was all about 21st-century skills. They used their critical thinking skills, they communicated with one another on many collaborative projects, and they were very creative.

Interesting Turn of Events

As the week went on I received an email from the New York State Education Department (SED). It was sent to their discussion list and focused on changes regarding the third- through eighth-grade ELA and Math exams. The document stated that our students would have to take four exams in two days.

Two of the ELA exams to be taken were 70 minutes each, which meant that our students would be taking exams for about 2.5 hours one day and almost 2 hours the next day. It almost seemed cruel to be sent these emails during a No Testing Week. It also seemed cruel to think that we focus so much on movement and hands-on learning in schools that we would make students sit for hours taking a paper-and-pencil exam.

However, all of that quickly changed when all public school principals and superintendents in New York State received three emails over two days. The following is what we received.

Dear Principal,
Please see the message below regarding the Grades 3–8 Testing Program. This message replaces the information you may be receiving as the system mails you an update from the Office of State Assessment. The information we have sent you is undergoing revision and is not accurate. Please wait for further information. Thank you and I apologize for the inconvenience (SED).

As I read down, it all became rather clear why there was a clarification email.

Dear Colleagues—earlier today a guidance document on the 2012 3–8 testing program was released prematurely. Please disregard the email and associated documents as the Department is still reviewing time lines, content, and structure, and will seek field input on these issues prior to final release.

We apologize for any confusion that this may have caused (SED).

To date we do not know if they realized the increase in testing time was borderline educational malpractice on our students, or they did not want us all to email and call SED. What we do know is that David Abrams, the assistant commissioner of the Office of Assessment Policy, Development and Administration, resigned the following day.

The state education department did increase the testing time for our students who are as young as seven, which is yet another symbol that our state education departments are going in the opposite direction of where schools need to go. High-stakes testing does not look at the whole child. It helps us understand only one type of learner.

It is up to our state and federal education departments to decide whether they want to continue down these dark days of education or move to the enlightenment where we are less focused on high-stakes testing, the wealthy publishers who create the tests, and the textbooks that go with them. We can be driven only by reliable data if we are truly going to help our students. Forcing our students to sit for six hours of exams over a two-week period does not result in good data.

Conclusion

For one week it felt like we won. Students and teachers were engaged in real learning, where every moment that surrounded them was an educational opportunity. They saw themselves in a different light. Our staff provided the freedom to have a week without testing where they got in touch with their creative side. I am extremely proud to work with a staff that created so many great learning experiences for kids.

The night of the author event we had parents and children walking around the building looking at all of the creative work that was created during the week. Our students were beaming as they pointed at their projects. Everyone took a walk through the 40-foot longhouse in the hallway and took pictures of it. There was a sense of pride in the building that, not only did we have a no-testing week, we surpassed any expectation we had of what the week would be like.

No Testing Week was a time to reflect on why we teach children and get a better understanding of what truly matters in education. Teaching all learners, regardless of whether they are a 1, 2, 3, or 4 on a high-stakes test, is why we entered the profession in the first place. The sad part of the event was how much publicity we garnered from having a week of no testing. It should not be a national story. A week without testing should be as common as the days of high-stakes testing has been for the past 10 to 15 years. We should be afforded academic freedom and be able to have more teachable moments within our school day.

Since our No Testing Week our school personnel and parents have had many conversations regarding how positive the week was for everyone in our school community. Many are talking about what projects to do the next time we decide to do a test-free week. It's exciting to see teachers and students engaged and happy. It's wonderful to hear from parents that their children loved the week, because that is what it is all about: our students.

Education in the United States is going in the wrong direction. We are forced to be driven by data that we do not control, from education departments at the state and federal levels that we no longer trust. Education is less about building relationships and creativity and more about proving that schools are failing, when we know they are not.

Things to keep in mind when planning a No Testing Week:

- Set up multiple meetings with staff.
- Communicate with parents. Many of them feel schools test too much and will support you.
- Try to plan one whole-school event a day, even if that means that the students will not have to leave their classroom (e.g., announcements, videos).

- If your school has a Scholastic Book Fair, plan your No Testing Week during that week. Bring in an author!
- Make sure you clearly articulate your goals for the week.
- A Week without Testing: A Whole Child Experience.

References

Markow, D., & Pieters, A. (2012, March). Metlife Survey of the American Teacher: Teachers, Parents and the Economy. New York: MetLife. Retrieved from *https://www.metlife.com/teacher-survey*

Rothstein, R. (2004). *Class and schools: Using social, economic, and educational reform to close the black-white achievement gap*. New York, NY: Teachers College Press.

Sadker, D., & Zittleman, K. (2004). Test anxiety: Are students failing tests—Or are tests failing students? *Phi Delta Kappan, 85*(10), 740–751.

Sharratt, L., & Fullan, M. (2012). *Putting faces on data.* Thousand Oaks, CA: Corwin.

FOURTEEN

Journey into Ungrading

Hadley J. Ferguson

It is often easy to identify the beginning of an adventure, but where that journey will take you is usually a mystery. That was certainly the case with my adventure into ungrading and using portfolios for assessment. There have been many unexpected twists and turns in the road, unanticipated challenges as well as significant and rewarding successes. When I asked my administration if I could teach an ungraded class, I knew that I was stepping away from the security of my established practice and into a place where all of my skills and knowledge would have to be applied in fresh ways. A new adventure was truly starting. I asked for and was given permission to teach the only ungraded class in my school. The school was in a time of transition, and teachers had been challenged to experiment with the best strategies for meeting the changing needs of 21st-century students. My class, seventh-grade history, became a place where learning took place within a new set of standards and expectations. While there were a wide variety of assignments and assessments, none of them was going to end in a grade.

Over the course of my career, I have moved from believing that bad grades were motivational to seeing them as tools that often reprimand and demoralize students. I have rarely seen the response to a bad grade be energy and engagement. When that was the reaction, when the negative feedback led to positive action, it almost exclusively came from students who regularly succeeded and believed in themselves as students and learners. The bad grade upset them and then led to action to make it go away. They believed that their efforts could bring about change, having the confidence and necessary tools to take on the challenge that came from their initial failure. For many students, however, the bad grade was simply one in a long line of poor grades that confirmed what they had already come to believe: that they were not as good as their peers, that they couldn't do what other students did, that their life was and would continue to be one of failure. They had no interest in comments written in the margin to help them make their next attempt better. The grade said all that they needed to know about themselves and led to discouragement and inaction.

Teaching outside the security of a system of grades that everyone understood was initially intimidating. Classrooms with grades were the familiar shore that everyone knew. We had all gone through those classrooms; we knew the rules and the roles that they assigned to every one of us. While some of us might be uncomfortable with evaluating students using a letter system, we understand how it works: An A means you are safe and smart, and an F means your work and, by association, you have been judged a failure. Letting go of that system and creating a new one forced the members of the education community to change and grow. From parents to the administration, from me as the teacher to the students, we all moved into the unknown.

In an ungraded classroom, the role of the teacher is the first one to shift. I was moving away from seeing my role as being the single authority in the room, the person who knew what good work looked like and passed judgment based on my knowledge and experience. Historically, the ability to give out a grade gave power to teachers. It was the constant threat that kept control in the classroom.

"If you don't take these notes, you'll get a bad grade."

"If you don't behave in class, you will fail this class."

When a teacher lets go of the power, it can be scary. By letting go, however, the source of energy and authority in the room shifts. The door is opened for a partnership of learning, one where both the teacher and the student have control over the process. I began to see my role as being more like a master craftsperson who is passing on the knowledge and skills of learning to my apprentices. It was still my job to know the way and to support each student's growth, but a significant part of the process became about training each student to identify the steps in her learning as well as the challenges that he or she encountered. I then had to provide the tools, taking away the mystery that can be the learning process, and support the student's understanding. My goal was more than to impart history content to each student; it was to create an independent learner.

Getting Started

Before school started, I wrestled with how to introduce this class to my students. I pondered what I wanted for my students:

What kind of students did I want for them to become?

What skills did each of them need to grow and eventually master the class?

What ideas and skills were the most important for them to acquire?

I wanted to be sure that I knew where this journey was going. Early on in my thinking, it became clear that for this to be successful in the way that I wanted, the students were going to have to learn how to effectively assess the work they were doing. They needed to be able to identify the skills they were practicing and then articulate what they did well and where they needed more practice. The challenge was finding ways to add that to the content of the class without destroying too much of the original content. I organized a list of the skills and main ideas that I wanted for the class, looking for ways to include minilessons on self-assessment.

Early in the school year, at Back to School Night, the parents needed to understand what was happening in the class. When the grades are removed, communication with parents becomes very important. My administrator introduced the idea to the parents so that it didn't look like this crazy, lone teacher had decided to change the rules. It was important to let the parents know that the lack of grades did not mean a lack of learning or assessment. It simply meant a different way of reporting the learning that happened to both the students and the parents. We wanted to make it clear that the students would still have work to do and if that work wasn't done or didn't move toward mastery, I would respond to that. I shared examples of the types of assignments that were going to be given and examples of how I would assess them and teach the students to assess themselves. I presented lists of skills that would be learned as well as some of the major ideas we would be considering. I explained that if their child was struggling in the class, the response would match the level of concern, ranging from meetings with the student to contacting the parent.

I set up a class website to highlight the work of the class and keep the parents informed. I took a lot of pictures of my students as they worked at various tasks, making sure that every student was in them. I published them on the class website with explanations of the work being done. I posted assignment sheets for significant projects along with the lists of skills to see the direction that we were headed. It took away much of the mystery from it being ungraded. They saw students reading, working together, writing, and creating. They could learn about projects and activities, if they were interested or concerned.

The Beginning with Students

At the beginning of this school year, when I told my students that in this class, they would not be given grades, I saw a look in their eyes that reminded me of my first time going whitewater rafting. A young college kid told us what to do, and I remember thinking that I was crazy to trust my entire family to this young man. He seemed perfectly nice, but did he really know what he was doing? That was clearly how my students felt when I told them they would not be assessed by a grading system.

"No grades! But how will we know if we are doing it right?"

They knew the rules were changing and weren't sure they liked it. While they might not like getting grades, they knew what school was like with them. School was just another place where the adults in their lives were in control: their parents, who wanted them to get good grades; and their teachers, who evaluated and assessed them at every turn. It was their job to perform: follow the rules, do what they were told, and hope that the evaluations proved them worthy. Now I was asking them to leave a place where they knew "how to do school."

"How will my parents know if I am a good student?"

They were programmed to work for the outside evaluation, people "in the know," who told them if they had learned or not. It was the adults who had the power to establish whether or not they are good students, if they were truly working hard enough. The motivation had nothing to do with intrinsic learning and everything to do with adult affirmation.

As I spoke to them, it was clear that they thought that if there were no grades, I was somehow abandoning them, leaving them on a deserted island, a *Lord of the Flies* kind of experience where they would have to find their own way.

"I don't know how to learn history!"

At this point, I talked about the skills and ideas of the class that I had outlined. I showed them the list and explained a bit about each skill and area of study. This conversation served two purposes: one, they saw right away where we were going; two, they understood that I knew the route. I reassured them that I wasn't going anywhere. My job was still to teach history and to teach them to be the strongest possible learners that they could be. I was going to show them ways to become better readers and writers. I was going to provide times for them to practice new skills and develop their thinking. The class was just going to be more about them in the midst of the history content. It was clear that they were not at all sure that they wanted to become part of this process of reflecting and evaluating the work that they did. It clearly felt scary and left them insecure.

At this point it was time to show them that they already knew a lot more about themselves as students and learners than they were usually given a chance to say. While they were still shaking their heads with worry, I set them to their first task. Using Peter Pappas's "Bank Robbery" activity (http://www.edteck.com/rigor/lessons/detective/clues2.pdf), I had them solve the mystery from a series of clues that they are given. It is a great critical thinking activity and a perfect set-up for a history class, because it got them to use two important history skills: asking questions and making categories.

They worked in groups of three, and I wandered around, offering encouragement and asking questions. When they thought that they had the solution, they explained the events to me to verify that they successfully unraveled the puzzle. When the groups had finished, we brainstormed a list of skills and strategies that had been used: listening to each other, taking notes, organizing, collaborating, etc. I then had them each write a reflection that answered the following questions: What did they do well in solving the mystery? What was challenging about the activity? What would they do differently if they had to do it again? These are the same questions that I continued to use throughout the year as guides for reflection. They are broad enough to allow for each student to talk about her work, but they don't limit the thinking. They quickly began writing, many of them nodding and smiling as they wrote.

I had them share some of their thoughts after they finished writing, and the whole class got excited as students began to understand that they did know a lot about themselves and how they worked. Their enthusiasm grew as they saw that they would have power to speak about themselves. Through this low-stakes activity, they saw that they did indeed know a lot about themselves, their strengths, and their challenges. They understood the impact of being listened to or being interrupted, whether they collaborated well or they didn't. They could talk about whether or not they helped organize information or if they simply got frustrated. They knew who they were as participants and learners in the activity and how they had participated; they just needed to be asked to talk about it.

For the rest of the year, I started each lesson or short unit by introducing the skills and main ideas around which the class would be focused. Sometimes, I presented that list at the beginning of the class and sometimes, as in the Bank Robbery activity, we generated it together at the end. Those lists became the foundation for the reflections and were key to involving the students in their own learning.

It was important for me to have an ongoing dialogue with each student about the learning in the class. Because all of the students in the class have laptops, I used Google Docs to comment on every piece of work and to encourage the next steps. It was easy to reference past work and interact around the work. This dialogue between us was an essential part of the class. I read all of their work and then left them comments, identifying the areas of strength and pointing them toward ways to improve on what they had done. I made sure that for each piece of work there was at least one positive comment and then one next-step comment. With students for whom a conversation with the teacher was intimidating, this provided an excellent means of supporting each student's understanding of the next steps required.

I wanted the dialogue to feel safe and encouraging. I did not want to overwhelm them with every aspect of what needed to improve, just to show each one where to head next. Without the ability to slap on a grade, it was sometimes tempting to use the comments to point out every mistake and failure. The paper or document with red ink all over it was no more supportive of student learning than the C- would have been. The goal always had to be to empower and encourage.

I tried to make sure I gave the students time during class to read my comments and reply back. I wanted them to know that this dialogue was important and so I had to give time to it. It didn't take much time away from class and reaffirmed the idea that their work had value. In the past, in the world of grades, when an assignment was done, all that was left was the grade. The assignment itself had no real significance, except as a vehicle to a grade. While I wasn't able to do it for every assignment, when the skills or thinking was important, I made time for the students to review their initial work and redo the areas where they were struggling, based on the dialogue that we had in the Google document. I wanted the students to see that reviewing and rethinking past work was important, that each assignment served a purpose.

One of the unexpected challenges of the class was striking a balance between a world of no grades and a learning environment where there were still facts to be learned and ideas to be discussed. Even in an ungraded class, there are answers that are right and there are answers that are wrong. When my students bumped into this, it was a challenging lesson for us—for me because I hadn't anticipated the impact it would have and therefore hadn't prepared them well enough for what happened. Before this challenge, the students were enjoying the sense of freedom that not having grades gave them. They weren't threatened when presented with new areas of learning or new skills, because it felt like they would be able to figure out how to master them without a sense of failure. They weren't stressed by the need to perform in a set way. The energy and confusion that came when they had to master a new skill, not just simply practice and play around with it, was very strong.

One of the skills that I wanted my students to learn was memorization, which I use to teach students about how they learn. It is one of the first parts of my process for teaching them how their individual brain works the most effectively. I set up a lesson to teach them the steps for writing an effective analytical paragraph and used this as the basis of my first memorization challenge for them. I wrote the steps on the whiteboard, discussing each step after I wrote it down. They asked questions and made connections with other writing they had done.

After we had finished discussing the process, we considered ways to memorize the steps. They each shared their best ideas for strategies, and then we talked about some others: creating songs, doing a dance, or writing it in the air. We practiced some of the strategies, and then they practiced them in groups. They experimented with a wide variety of methods, and in the process they began to learn the information. I had them create a Google Doc on Memorization to keep track of their best strategies for learning information. They recorded the approaches they had tried and whether or not they thought they were effective. It was exciting for them to recognize what worked for them and how they were different from their peers.

About a week later, I gave them another memorization challenge, learning the aspects of an empire, again information over which I wanted them to have control. I presented the information and then they worked in small groups and independently to learn it. Together we decided on a day by which time they should have learned the material. They set up a study plan in their Google Docs, and then we went on to other activities. When they came into class on the set day, I had them write out the aspects they had learned. After they were finished, we went over the answers together. Because the goal was mastery, I had them evaluate their answers. I told them to choose how to identify the ones that were completely correct—with a star, a check, a point—and then the ones that were only partially correct, and finally the ones that were incorrect. When we finished going over the answers, they immediately added up the number of correct responses that they had and graded themselves.

They suddenly were greatly agitated. They felt judged by the number of stars they did or did not have.

"This isn't fair. This class doesn't get grades. How can you give us a quiz?"

"Are you going to tell our parents?"

"Do we have to turn this in?"

Suddenly when faced with a number that showed how many were correct, a number that they quickly translated into a grade, they were angry and confused. For many of them, especially those who struggle in school, a poor score seemed to immediately mean parent contact. They were now in a scary place, one that they, interestingly, had created themselves. The information was not going into a grade book. Writing down the steps was for them to see whether or not they had been successful in their learning and to identify what they needed to do next. If they had been, then they were ready to go on to the next step. If they hadn't, then there was a need for new strategies. The decision to see it as a quiz was something they created, though it was something for which I should have anticipated and prepared them.

As we talked and I probed a bit, what I discovered was that because they weren't receiving a grade, they had felt comfortable with actually not doing the work that they would have done for a "real" quiz. They relaxed, not into their learning, but away from it. It had been fine to simply read over their notes, because there was no heavy stick of a grade, but when they realized that they themselves could see their success or lack thereof, they didn't like it. They wished that they had followed their own plan, because then they would have had success. They began to see that they actually wanted to reach mastery. They liked to be successful and know the material; they didn't want to fail, with or without a grade. They wanted to learn. It was important to them.

While I could have orchestrated the process more effectively, they had gotten to an understanding that I wanted for them. They were starting to take on the responsibility for being accountable for their learning. At the end of the conversation, they wanted a chance to redo learning the information. Together we decided to take three more days, during which time they could come in and write out the aspects. Most of them had it memorized and completed by the next day, clearly showing pride in their commitment and accomplishment. It wasn't that they couldn't do it; it was just that they had misunderstood how to approach the work of the class.

Even when they began to take ownership for their learning, there was the ongoing challenge of being the only ungraded class in the school. The students had to constantly wrestle with the challenge of juggling their many graded classes with their ungraded class. If they had to make a choice between the graded quiz in one class and the ungraded work in mine, they chose, more often than not, to put their attention toward the graded one. In their overall world of school, the grades mattered, which meant that the ungraded one necessarily did not, or at least not as much; they were following the training on how to be successful in school that they knew. This was a hard place for all of us to be. I wanted my class to have value, and I wanted students to be engaged. If I let my students take an assessment multiple times to reach mastery and another teacher gave it only once, then the student seemed to have little or no choice on where to put effort.

At this point, I suggested that we build some routines into their studies to provide them with more structure for their learning. Middle school students work well within structure, so I wanted to help them establish some parameters. Grades often do that. The fear of a bad grade creates the drive to do the work. I wanted for us together to develop routines that would give them guidance without the fear.

The students immediately responded to the idea of creating their own class rules to add structure and security to their work. As we talked through which routines were important for them, it was clear that they were trying to find limits for themselves, so that other classes didn't take away from their time on History. It was less that they wanted to be disciplined than that they wanted rules that added value to the work of this class. It gave them more power to put aside another class to study for History.

Their first idea surprised me: They wanted quizzes and even tests periodically, because it gave them a sense of accomplishment to know that they had control over a body of information. They wanted to be able to retake any one in which they didn't feel they had showed mastery. They only wanted two more chances; then they felt it should be time to move on. They wanted some homework, but not too much. If they didn't get it done, they wanted another day with no consequences. They thought that their parents should be contacted only after three homework assignments had not been turned in or after three tries on the test didn't show mastery. As they talked, it was clear that they wanted to hold themselves accountable in ways that would push them to do their best work over time. Some of this clearly came out of the structure of the school that they are used to, but some of it also came from them trying to find ways to establish good learning routines.

One way to provide the next level of challenge was through ongoing feedback for the students. Without the standard of "I got a B" to rely on, students needed to have a clear sense of where they stood in their growth and learning. I first used rubrics to help them see what the goal of a project or activity was and to understand what some of the aspects of excellence might be. I quickly found that I was no more able to fit an individual student into the boxes of a rubric than into the boxes of a grade. The part of the rubric that helped my students was the list of skills that they were going to practice through each activity. As the year progressed, the students were able to generate this list of skills, but initially I created it for them.

When a project or assignment was completed, the first step in the evaluation process was for each student to reflect on her work and engagement in learning. Using the list of skills, the student described the work, thinking about successes, challenges, and what she would do differently. I also wrote an assessment of the work that had been done, providing feedback on what the next step for each student should be. Over the year, the students learned to develop their own list of skills based on the work that they wanted to accomplish. They identified the types of learning in which they would be engaged and took increasing ownership over the process.

One of the goals of my ungraded class was to develop strong, independent thinkers. I wanted class time to allow for investigation with lots of opportunities for students to become curious. I did that with primary source work, where they were seeking to understand the past through the documents and images of the time. It was important that learning was done beyond the world of deskwork and worksheets. It required activities that pushed them gently, but firmly, into new ways of learning. Each challenge needed to lead to the next one, a little harder but still within reach.

Reporting from an Ungraded Class

At the end of the first trimester, it was time for the students to put together portfolios to present their work to their parents, in lieu of a grade and comment from me. The students had grown more adept at identifying the skills that they were learning as well as their personal strengths and challenges. They had moved away, not completely but to a large extent, from expecting that I was going to tell them, through a grade, what kind of student they were, good or bad. They had begun to develop independence around their learning, which was very exciting.

I wanted to make sure that the process of creating their portfolio was organized in a way that allowed them to develop a complete picture of where they had started in September and where they had moved to in three months, and then in six months, and then in nine months. Each student created a Google Doc that was titled "Portfolio Planning." We generated a list of skills that they had been working on as well as a list of the activities that they had done to learn new ideas and to practice their skills. They came up with a great list, everything from new memorization strategies to taking walks to get the blood circulating, from how to annotate a history text to writing an effective paragraph. They demonstrated a clear grasp of the work that they had done, showing me that the consistent naming of the skills was having an impact on their understanding of the work that they were doing.

We then organized the skills that they had generated into five categories:

1. Studentship (which included preparation for class, class participation, and collaboration)
2. Thinking and Learning (memorization, organization, learning styles: strengths and challenges)
3. Reading
4. Writing
5. Group work

The following are the instructions that I gave to the students to direct their reflections:

1st Trimester Portfolio Guidelines

The goal of your portfolio is to show your growth as a student in history over the fall trimester. You will think about yourself in terms of different types of skills. Here are the major categories:

Participation and preparation;

Thinking and learning;

Reading;

Writing.

You will be selecting pieces of your work to illustrate your skills.

Required Pages:

1. Yourself as a student of history in class:

Participation during group discussions (asking and answering questions)

Collaboration with partners

Studentship

Preparation for class (materials, homework)

Engagement in class during assignments

2. Your reading

Identifying main ideas

Actively engaging the text

Annotating

Identifying significant ideas

Asking questions

3. Yourself as a writer of history

Follow steps to a paragraph

Translate your argument into a topic sentence or thesis statement

Support your argument with evidence

4. Yourself as a thinker and learner

Ability to learn information

Ability to recall information

Ability to make connections between information

Recognize patterns

Develop an argument

Put information into your own words

5. Goals for the coming trimester: After reviewing your work on these different skills, identify one to two specific skills that you would like to focus on in the coming trimester.

At this point, I returned student work from the first week or two of school that I had saved just for this purpose and had them compare it with their current achievements. I had them go back to the six to seven reflections that they had written over the trimester. They reread each reflection from earlier activities and then responded to the different aspects of the worksheet. They first identified what they had been like at the beginning of the trimester and then how they had changed or grown in each category. The students also wrote about what they felt comfortable with and where they felt challenged. Finally, they set goals for themselves for the coming trimester.

I then went into each student's "Portfolio Planning" document and responded to the writing.

The following is an example of this work: The student comment is in **bold;** my comments are in *italic*:

1. Yourself as a student of history in class:
   a. Participation during group discussions (asking and answering questions): **Could ask more questions and answer more.**
   b. Collaboration with partners: **Work fine with partners. Help if needed. Sometimes get distracted or get off topic. Blow through class work more than I should, hard to pay attention without fidgeting but shouldn't have the balls** [There are squishy balls that I keep on my desk for those that are helped by using them] *Can you think of something that would help you to slow down and focus on the task at hand? What distracts you? Do the balls draw your attention away? Would writing notes help you focus?*
   c. Responsibility for yourself: **Bring books and papers and occasionally forget to bring a material and always forget to bring pencils or pen.** *What kind of goal do you want to set for yourself here?*

In some cases, I left comments in the margins (see Table 14.1).

Table 14.1. Sample Comments

How did you study for this?	Give an example or two of how you helped your group.	Would reviewing the work before class give you some confidence in what you have to offer to the class discussion?

The next step was for them to respond to my prompts, editing and revising their initial work. When that was completed, they turned their notes into short descriptions of themselves for each skill. Each student wrote three to five sentences for each of the five categories that included strengths, challenges, and at least one goal for the coming trimester. I again reviewed each student's writing and made suggestions for revisions. It definitely took more time than tabulating up the average from a series of numbers, but the results were wonderful, thoughtful presentations of what each student knew about herself.

It was now time to create the actual portfolio. We talked about ways to make the portfolio interesting for the parents and decided that on each page, there would be student writing with at least one visual aspect. The students chose different possibilities for their visual: a photograph from my class collection, a scan of her work, a screen shot. The most amusing moment in the process came when students decided that they needed a series of "reenactment" shots to show parts of class life that were not documented: hand-raising to answer a question or taking notes. I took photographs from all angles to make sure that each student had at least one "reenactment" photo to add to her pages. They then took screen shots of the lists of documents in their Google History collection and of various mind maps and documents that highlighted their learning. They scanned examples of their first attempts at annotating and their last. Using Haiku ePortfolio, they organized their images and writing onto four pages. They then created an "Introduction to My Portfolio" movie, 20–30 seconds, in Photo Booth for their parents, explaining what was in the portfolio and its goal.

I then reviewed each portfolio—time consuming, but exciting. They demonstrated such control over their learning! I decided that I wanted to provide them with editing comments, but not make the changes myself. The goal was for them to speak for themselves, not for me to speak for them. I knew, however, that they wanted their portfolios to be well done, so I created a Word document for each of them. I cut and pasted any writing that needed review or revision into the document. Depending on the student, I provided direct instruction. Also, if there were places where the language was confusing or needed further explanation, I highlighted it and posed a question to direct the editing process.

I gave them time in class to do the final edits. When each student decided that her portfolio was complete, it was time to hit the "Publish" tab and make it live. It was an exciting, but a nervous, moment for each one.

"Are you sure? Should I do it?"

My response was to put it back to the student.

"Is it ready? If it is, then go!"

Individually and sometimes in small groups, they counted down, "3, 2, 1, Publish!" and their portfolios were open to the world.

One student turned to me as soon as she had pushed "Publish" and said, "Now that our portfolios are done, will you give us our grades?"

I just had to smile! Even though she had never received a grade from me, she was convinced that there had to have been a grade book, hidden away somewhere, and that now, I was going to pull it out and let them know how they *really* did. It was hard for her to believe that I had actually meant it when I said it was ungraded, not un-assessed.

For the report cards that were being mailed, I wrote the following blurb to go in the history section of the report card, reminding the parents about the aspects of the class and directing them to the online portfolio:

> In history this trimester, your daughter has grown in her understanding of herself as a student. She has learned to identify the skills that she is practicing and to monitor the level of her mastery. With my guidance, she has reflected on her learning at each stage, and has then set new goals for herself. Each girl built a portfolio to share how she sees herself as a student of history. All of her work was closely monitored throughout the trimester. While I edited every website and gave each girl notes for revisions, some errors have been allowed to remain. One

goal is for the girls to improve their proofreading over the course of the year, and when they look back on these, see their growth.

The students published their first portfolios right before Thanksgiving weekend. I confess that I held my breath that weekend; I wasn't sure what the reaction was going to be. I knew that the portfolios demonstrated an incredible amount of work and thought on each student's part. Each of them had managed to communicate how hard she had worked in history and where she had grown, but it still wasn't a grade. It wasn't a standard form of assessment, one that came from one authority, the teacher, to the other authorities, parents and administration. Each portfolio was the student speaking for herself, explaining her learning and her challenges.

What followed was silence! No phone calls, no emails, nothing! I wasn't sure whether to be pleased or to wait for some other shoe to drop. When the week started, I investigated, asking the students how their parents had responded to their portfolios. While the responses varied, it was clear that the silence over the weekend had two sources: The first was that the parents had simply not had enough time to sit down and read all that their children had written. Rather than having to digest a grade with a short comment, there was an entire website: pictures, documents, and commentary. It was far more than they usually received. One student said that her father had responded to seeing the portfolio by saying that he still wanted a grade. She said to me, "Don't worry! I explained it all to him."

The second reason for the silence was that they were impressed with how much their child understood her own work. She could talk about it and answer their questions. There was no need for me to do it. The conversation no longer needed to be simply with the teacher; their child was more than capable of speaking for and about her learning. She knew her strengths and her challenges. She had set goals for herself moving forward and had organized a plan to attain them.

The portfolios at this point became the students' evolving record of the learning that was happening in the classroom. After taking the time to create the portfolio, the students added to it more organically. We followed a pattern where I introduced new skills in minilessons, similar to the memorization lesson. We practiced the skill or new idea together, and we discussed what they thought the challenges for them would be. Because the portfolio wasn't organized around content units, but instead was around aspects of their learning, I set up the lessons so that they clearly connected to the pages in the portfolio. They created new pages for the winter and spring trimesters and added subpages of skills to those. There was always time at the end of a series of activities to write about their successes and new goals in reflections, adding images to more fully present their thinking.

Adding to the portfolio became part of the reflection process. It was where they showed what they knew about their growth as students. Google Docs continued to be the primary place where I left them comments, and we had dialogues about their work. I periodically reviewed the portfolios, making suggestions for what they might want to add.

When it was time for the second and third trimesters to end, the administration asked me to also write comments for the students, so that the parents heard from the teacher more directly. I did not want to take away from the power of the student voice, so I created a blend of my voice and theirs in the comment. I took them through a similar process in identifying how they had grown over the trimester. This time, they had their portfolio to refer to. I then had them write 7–10 sentences about themselves in History, focusing again on strengths, challenges, and goals. Then I used parts of their writing within their comment to show how well each understood her learning. The students and their parents heard their voice and their understanding of themselves and saw my affirmation of the student's self-assessment.

> Margaret's energy and excitement in history were often contagious. On the days where she felt secure with the material, she was an active participant in the conversation, willing to share her ideas and to ask questions to verify her understanding. She enthusiastically wrestled with the questions of history. As Margaret wrote about herself, "I am an enthusiastic and engaged history student. I think that I did a lot better with class participation because I know how this class works now, as opposed to before, so I feel like I can contribute more without feeling weird or nervous. I also feel that I am more committed to my work because I want to do well." When working in a small group, Margaret presented her thinking and collaborated with her partners to achieve the group's goal. When reading, Margaret worked on developing her annotation skills. Initially, she read to identify the main ideas and then reread for the significant facts connected to them. Margaret set goals for her reading: "I will also try to read slower and not try to rush just because one person is ahead of me, I'd rather understand the information and read slower than skim the information and read as fast as I can. During the times that we read, if we are allowed to spread out, I will try not to sit close to a friend (one who might distract me) so I can get as much done as possible. I think that when I annotate I could be a little bit neater, so that when I go back over the packet, it doesn't just look like a whole bunch of scribbles and lines. I want it to look like I care, so that I know what facts to study and what places to look back over." Margaret had clear success in developing her research skills over the trimester. She carefully identified significant facts from the resources and recorded them. Margaret found it difficult to paraphrase direct quotations, and this is an area where she should continue to work. The greatest challenge for Margaret came when she had not fully prepared for class. On those days, she was more reserved in class, less willing to test her ideas and participate in the class conversation. As Margaret said, "I will go into the next term feeling a lot of confidence and I will try to finish off the year with great ending. I will try

> to work harder than I have before." A wonderful goal! I look forward to working with Margaret for the rest of the year.

Margaret (not her real name) was not a student who easily had success. She struggled to complete assignments and often spent a large amount of energy trying to be as invisible as possible to her teachers. She was so excited to reach a place where she could name her strengths and set personal goals for herself. The areas in which she struggled didn't beat her down; they were now starting points for reaching toward success. She learned to take more chances and as a result began to grow as an active and engaged member of the class.

Before the reports were sent home each trimester, the students took time to review what they had added to their portfolios. They often asked each other for suggestions and help in making them more interesting. Some of them added videos; some added just pictures. Most of them had multiple scans of their work and screen shots from their Google Docs. For each one, they wanted to make sure that they had a full representation of their work and learning, one that their parents would learn from and find engaging. They were proud of what they had learned about themselves and they wanted to create a forum that fully presented it.

All students want to learn; they want to be smart and succeed, but we have created a system that deprives most of them of the joy of learning. For the students who succeed in school, they live under the constant pressure and stress of losing control, of not getting the A. For the ones who struggle, they know that they are failures and refuse to try one more time. And for all of those in the middle, they have lost much of their curiosity in the millrace of attending school.

Teaching an ungraded class required hard work and commitment from the students and me. It pushed everyone away from what we knew already and into a world of new learning. There were endless new horizons and adventures, ones that we all learned to tackle and eventually master, that replaced the security of just teaching and learning right answers. There was no passivity in the teaching or the learning. While it required effort and engagement, it provided the students with a safe environment in which to work. It gave them the freedom to experiment. If they didn't have success, there was not condemnation, only encouragement to try again in another way. In the end, as one student said to me, "I work really hard in school to get good grades. In this class, I worked really hard because I was learning." The students left knowing themselves and how they learned, their strengths and their challenges. They also knew some strategies for handling the more challenging aspects of learning. In every class, they were ready to tackle new problems with greater self-confidence. They learned it in history, but they carried it forward with them.

|| FIFTEEN

Moving Beyond "Parents Just Want to Know the Grade!" *

Jim Webber and Maja Wilson

Occasionally, someone has the nerve to suggest that grades are overrated, that a focus on them is detrimental, and that everyone might be happier and learn more if we deemphasized or got rid of them completely. A widely discussed article in *Inside Higher Education* (Jaschik, 2010) described Cathy Davidson's efforts to "get out of the grading business." In her English classes at Duke University, students held regular meetings to decide if their work was acceptable or needed revision. Davidson gave no grades—only descriptive feedback. At the end of the experiment, Davidson declared, "It was spectacular. . . . It would take a lot to get me back to a conventional form of grading ever again."

The success of Davidson's class experiment and its implied indictment of grades isn't really news. Since the publication of *Punished by Rewards*, Alfie Kohn (1993) has been telling educators what psychological researchers already know: Grades and other extrinsic motivators not only make students less likely to choose challenging tasks, but they also reduce the quality of their thinking and their interest in learning. More recently, Daniel Pink (2009) has taken up Kohn's mantra in *Drive*: Incentives such as grades may work for short-term compliance on simple tasks, but they undermine intrinsic motivation and performance on complex tasks. Writing teachers who are irritated by students' plaintive "Tell me exactly what to change for three more points on this essay!" understand this without reading the research.

Still, the research accumulates: A study (Pulfrey, Buchs, & Butera, 2011) demonstrated that when students anticipate grades on papers (with or without com-

* Adapted from article originally published in *Phi Delta Kappan* (September 2012).

ments), they become more likely to avoid difficult work than when they anticipate teacher comments without grades. This finding complements Ruth Butler's (1987) study showing that grades (with or without comments) lead to lower levels of intrinsic motivation and creativity. But suggest that we act on this research—by deemphasizing or replacing grades in the classroom—and even sympathetic teachers conjure up parent protests: "I'd be the first to get rid of grades and just do writing conferences and narrative feedback! But parents just want to see the grade!"

Lisa Riffles's classroom would prove the perfect place to put the parent excuse to the test. Lisa, a well-known literacy educator, has been quietly deemphasizing grades in her eighth-grade English classes for the past 30 years. She is required to give semester grades, but several times a year, in lieu of intermediary grades, she sends home a binder with drafts and revisions, her written responses to this work, and a letter to parents describing each child's progress and challenges. Lisa asks parents to read and discuss this work with their children.

But even Lisa was starting to wonder if parents valued these artifacts and descriptions of learning. She confided to us (Jim Webber and Maja Wilson) in 2009 that she'd spent hours writing letters the year before and had heard only four responses from parents: all of them inquiries about grades. We (Jim and Maja) decided to investigate. Did parents of Lisa's students *really* just want to know the grade? And, did Lisa's practice deny parents what they wanted?

What Do Parents Think, Believe, and Want?

A survey might have quickly gauged parental attitudes. But Lisa's leeriness of reducing her knowledge of students' learning to numbers warned us that a conventional questionnaire might invite similarly reductive expressions of parents' concerns. Lisa believed in the power of description and narrative. In that spirit, we wanted to give parents the opportunity to tell their own stories.

We invited all the parents of Lisa's students to participate in hour-long interviews about home-school communication. We interviewed the seven parents who responded to our request for participants. While the number of interviews we conducted was small, the parents we interviewed expressed a full range of positions on the value of grading: from one parent who envied teachers' ability to rank students against each other to a couple who enjoined teachers to spend less time entering grades into the Web-based grade sheet and more time interacting with their daughter. While every school community has its own character, we think the range of parent concerns we encountered is likely to be present in most communities.

We encourage educators who invoke parents' preferences to consider what we found: What parents *want* is more complex than what they *say* in brief interactions with teachers and administrators. Even parents who value grades and other forms of ranking wish to engage in ways not ultimately satisfied by grades.

How Edline Influenced Conversations about Parent-Teacher Communication

Lisa's district had just adopted Edline, a Web-based program that posts grades online. At first, teachers were encouraged to post grades weekly, but this encouragement would change to a mandate the following year. Understanding Edline's role in shaping teacher and parent definitions and expectations of communication would become central to our understanding of what parents want.

Programs such as Edline (Schools Open, Grademark, etc.) have spread quickly across the country, claiming to help schools "raise student achievement by harnessing the power of parental involvement, supporting teachers and engaging the learning community" (Edline, 2011). These programs pose a particular challenge to teachers like Lisa, who believe that grades are insufficient to communicate the richness of their understanding about students' progress as readers and writers. They replace questions such as, "What's the best way to represent and communicate what I know?" with questions such as, "How many grades should I give per week?"* and, "How often will I be expected to upload these scores onto Edline?"

Despite their promises, these programs make problematic assumptions about learning, representations of learning, and communication. First, they assume that if everyone simply focuses more on scores, *learning* increases, an assumption clearly contradicted by the research. Secondly, they assume that what teachers know about their students' progress can be accurately represented by a series of scores, an assumption Lisa does not share. And, most importantly for the purposes of our investigation, these programs redefine *communication* as the *reporting of scores*, and assume that these reports satisfy parents' desire for communication.

Parent Interviews

We began our interviews by asking parents to describe their own literacy and school histories. Then, we asked them to tell stories illustrating key moments in their children's school experiences. Next, we solicited examples of communication between the school and home. Then, we asked parents to discuss the relative or cumulative value of all of these forms of communication.

We'll introduce you here to three parents: Michael's mother, Sandra, and Adam's parents, Alan and Lydia.** These conversations illustrate what we found across all seven interviews; parents' positions on the value of grades were complex and often shifted throughout the course of the interview. And every parent—even

* In January 2010, a discussion thread with the title, "How Many Grades/Scores Per Week?" generated 189 comments on the English Companion Ning, an online discussion forum for English teachers. Only a few posters attempted to question the premise of the thread, and these posts were generally ignored. In fact, one poster asserted that teachers who give fewer grades are generally teaching very little.

** Names have been changed.

Sandra, who wanted to know how better to rank her children against others—was adamant in the end that access to student work and teacher comments, conversations with the teacher, and narrative descriptions of learning were most important.

What Sandra Wants

Michael's mother, Sandra, repeatedly identified herself as a math-science person. She imagined that her eighth-grade son Michael would be an engineer someday. "Getting him to write," she admitted, "is like pulling teeth." Still, she described him as "a big reader" and talked at length about the elementary school's role in fostering a love and habit of reading in him by requiring that all the children read 20 or 30 minutes every night.

Sandra stressed the importance of *objective* criteria and evaluation, attributing this emphasis to her background in science and math. The first story she told about Michael's literacy history involved multiple-choice reading tests she administered herself at home—when Michael was in second grade. While she was pleased that his teacher assigned independent reading as homework, she felt the need to supplement this reading with online multiple-choice reading comprehension tests whenever Michael finished a book.

Sandra expressed a strong desire to know how Michael compared to other students his age. She asserted, "I would love to be able to look at everyone from the class and see if the benchmark is with everyone else. I'm envious of the teachers in that because they know so much more about what is at that level developmentally."

Sandra's desire for ranking—a way of representing student achievement that lends itself toward grading—makes a perfect pro-grading sound bite. While Sandra didn't mention Edline specifically, we could imagine her featured on Edline's promotional materials, praising the program for permitting easy access to Michael's scores. A quick conversation with Sandra might lead a teacher or administrator to assert, "Parents just want to know the grade."

But this would miss most of the story. When asked to talk about the various ways that teachers had communicated with her about Michael's learning over the years, Sandra enthusiastically described the binder that Lisa sent home quarterly. "It feels different," she said, "to see the work than to see the grade."

Earlier in the interview, Sandra had talked about Michael's relationship with his second-grade teacher, Mrs. K, with whom he had "really clicked." Mrs. K asked the students to write a note about what they'd done during the day to give to parents at night. Sandra still has a folder of these notes, and she laughed as she recounted its contents: "He wrote on one of the notes . . . 'Mrs. K is a rogue.' She had taught the word to him, and then he turned it around on her. They had a kind of understanding, I think."

Sandra claimed that this assignment was the best form of teacher-parent communication because it allowed her to know what had happened in class every day *and* allowed her to see Michael's writing. We suspect that if we hadn't invited Sandra to tell stories about Michael's learning, she may not have discussed these notes at all, much less identified them as examples of parent-teacher communication. After all, they were an assignment that Michael had completed, not a conversation between parent and teacher. They included artifacts of his learning, indicating what he'd done, not judgments of how well he'd done it. The context of a conversation—which included the time to tell stories and make connections between them—made these notes available to Sandra as a desirable example of communication.

The complexity of Sandra's attitudes about grades and communication should give us pause when we assert that parents just want to see the grades. If Sandra—the most enthusiastic proponent of grades we interviewed—valued grades *less* than the notes Michael wrote about his day and the folders Lisa sent home with Michael's work and her comments, then we simply can't justify spending more time on grades and less time on other forms of assessment and communication on the basis of what parents want. There's much more to what parents want, it seems, than what they say in the few moments when they have the ear of teachers and administrators.

What Alan and Lydia Want—and How What They Want Changed through Conversation

For many years, Adam's parents—Alan and Lydia—had collected books for their children, and they recognized and encouraged Adam's writing. They explained that Adam enjoyed the Beat writers, and his own writings had begun to reflect the vocabulary, sentence structures, and tones of his favorite readings.

At the beginning of the interview, Alan wasn't sure that, by deemphasizing grades, Lisa was pushing Adam hard enough. Alan said carefully, "We suspect that he's not getting what he needs in school about correction." Alan explained that Adam's voracious reading habit might be causing a weakness that correction could fix. He described the problem:

> [Adam's writing is] stream-of-consciousness, [using] words just because he likes the sound of [them]. . . . Lisa is on to him, encouraging and saying some good things. But you look at his writing, and you see 1001 editing opportunities, and [his teacher] is not doing it.

Alan opens the portfolio of Adam's work and rummages around for something:

> Perfect example. Let me read you something [from a reflection on a paper Adam had written]: *this piece was written in the hopes that I could chronicle the final lives*

> *of a character whose social age I hadn't written about much lately. I had dialected youth in the novel I was and am writing and so along with just a general idea I wished to see if I could drop of the case of versatility.*

Alan laughs, grimaces, and looks up at us:

> So, [Lisa's] comment is, "What is dialected youth?" You know, okay. Good comment, but I'm looking at this sentence, [thinking the problem starts] with not even capitalizing the *t* in the first word along with all the other stuff that's going on in there. There's a lot to improve and you have to balance that with his thinking and the way he writes, the creativity of it.

Jim asked if Alan and Lydia tried to do this with Adam. They exchanged a glance and answered in unison: "It's a struggle." Lydia explained how she had to help Adam edit his prose:

> When we started reading his writing, I went right over and opened up *Strunk and White*—"use few words." [I said] "Adam, there's a beauty [in few words]" . . . we talked about haiku [but] I didn't get far in that.

Lydia and Alan knew from experience that their child's personality matters. As Alan and Lydia reflected on their experience with Adam, Lydia talked them both into a new understanding:

> When we were raising the children, there was a Think that said, *it doesn't matter what they read, just that they read*—cereal boxes, newspaper, anything. I sometimes wonder if there's that Think in writing: *it doesn't matter what they're writing, just as long as they're writing*. There's a time and place for everything—[we can] pull him in and get him more structured the more reading experience he has. I read a study that said that writers need tons of reading, if there's one necessary condition for writing—even more important than writing—it's reading. He is well on that track . . . I sometimes think this will work itself out the more informed a reader he becomes. Right now, he's absorbed, like listening to music. But then he'll become more . . .

Lydia trailed off and Alan interjected:

> That's a wacko sentence I read to you, but it is sort of typical. You can't understand him, but he's taking you somewhere, and it's going to be interesting, where he takes you.

In other words: Lisa's selective response to Adam's errors might not indicate ignorance of his needs, but a pedagogical decision. This conversation couldn't have taken place around a report card. It required samples of Adam's work and Lisa's comments. Descriptive assessment and artifacts of teaching and learning allow and depend on a substantial conversational process.

Communication and Conversation

The interviews suggested a central insight that teachers and administrators can apply when they consider communicating with parents. *Reporting* is not synonymous with *communication.* A better synonym might be *conversation.* Certainly, not all communication between parent and teacher needs to take place through conversation—as illustrated by the notes that Sandra still keeps. But schools should consider the characteristics of good conversation when developing and evaluating venues for communication between school and home. Otherwise, their view of communication will be stunted.

Conversations create space for the complex dynamics necessary for communication. In a good conversation, participants feel as if they're realizing things they'd only suspected before. A good conversation moves—it builds and bends back on itself, pauses, gathers steam, takes turns, plunges forward, and gathers itself into new understandings and connections. The full engagement of those involved is required, as participants give voice to their experience to create and complicate shared understandings. Determining what parents want without giving these dynamics a chance to play out is an exercise in speculation.

So, let's evaluate Edline's ability to create *conversation* between school and home. It provides no forum for parents or teachers to describe what they know about a child's struggles, interests, or achievements. It doesn't allow for the exchange of stories that parents or teachers might use to build careful generalizations about kids as learners—or to serve as a check against generalizations made with haste or even prejudice. It doesn't allow parents to participate at all, except as recipients or consumers of a report.

Many schools look to these programs to fulfill their communicative obligation. But at what cost? What Edline provides isn't really communication, and it lulls everyone into thinking that the impoverished view of communication it offers is all anyone has a right to expect. The programs may also contribute to an unhealthy obsession with grades. Edline's promotional materials include an article that proudly quotes a student's misgivings about the program, "On the downside, my mom's addicted to it." This addiction may not point to the inherent desirability of the information Edline provides. Instead, it may indicate how ultimately unsatisfying it is.

The goal of our conversations was to create space for satisfying communication. Our interviews suggested that what parents wanted was more complex than what they said, so we invited parents to explore the full range of their values and concerns. To encourage this kind of inquiry, we asked parents to describe their children's experiences with literacy *before* forwarding position statements on practice (e.g., do you think selective correction is responsible?). We found that inquiry into experience enabled parents to acknowledge the complexity of "what they want" as well as appreciate how teachers' practices supported their values. Moreover, we found parents willing to acknowledge the range of their values, even when some statements seemed to contradict each other.

Our approach sought to balance two central goals in teaching: inviting meaningful public engagement *and* building support for teachers' judgment. By framing parents' assessments of teachers' practice, we ensured that local conversations did not simply accept public commonplaces about assessment ("teachers aren't doing their jobs" or "my son/daughter isn't getting what he/she needs"). Instead, our conversations offered parents the opportunity to develop nuanced assessments of progressive assessment practice—the kind that make dialogue between educators and the public possible.

We understand that public engagement—like dialogue with parents—requires listening to criticism of teachers' judgment. This can be an uncomfortable experience and may lead many of us to change our practice preemptively to avoid conflict. But we don't believe teachers serve their profession *or* the public by accepting what parents *say* as evidence of what parents *want*. Rather, we recognize that public values are more capacious than public discourse about education. Parents and other publics want many things from teaching and learning, many of them aligned with progressive educators' values. Unless we consciously frame discussions with parents and the public, however, we squander the opportunity to build support for our practice. While the home-school relationships created by reporting may suggest that parents "just want to see the grade," we believe that conversation will reveal potential common ground between public values and our professional values.

Conclusion and Implications

Teachers' objections to emphasizing narrative, descriptive evaluation and deemphasizing grades cannot rest on uninformed claims about *what parents want*. As decades of research show, grades don't lead to deeper understandings, increased intellectual risk-taking, or better performance on complex tasks—and so conversations based around grades can't produce these results, either. Our interviews showed that parents want teachers who observe their children carefully, develop strong learning relationships, and communicate meaningfully. Our small study only illustrates, in richer detail, what a PDK/Kappan poll found: public school parents ranked examples of student work and teacher observations significantly higher than grades and test scores as the most accurate picture of student progress (Bushaw & Gallup, 2008, p. 17). Parents *are* telling us what they want: fewer grades, more description, and more shared artifacts of teaching and learning. It's time we listen.

References

Bushaw, W. J., & Gallup, A.M. (2008, September). Highlights of the 40th PDK/Gallup Poll; Americans speak out—Are educators and policy makers listening? *Phi Delta Kappan, 90*, 8–20. Retrieved from http://www.ncrbc.net/static/doc/40th_PDK-Gallup_Poll_Highlights.pdf

Butler, R. (1987). Task-involving and ego-involving properties of evaluation. *Journal of Educational Psychology, 79*, 474–482.

Davidson, C. (2009, July 26). How to crowdsource grading. [Web log post]. Retrieved from http://hastac.org/blogs/cathy-davidson/how-crowdsource-grading

Edline: A Blackboard Company. (2011). The Edline difference: Why Edline? Retrieved from http://www.edline.com/why-edline/the-edline-difference/

Jaschik, S. (2010, May 3). No grading, more learning. [Web log post]. Retrieved from http://www.insidehighered.com/news/2010/05/03/grading

Kohn, A. (1993). *Punished by rewards: The trouble with gold stars, incentive plans, A's, praise, and other bribes*. Boston: Houghton Mifflin.

Pink, D. (2009). *Drive: The surprising truth about what motivates us*. New York: Riverhead.

Pulfrey, C., Buchs, C., & Butera, F. (2011). Why grades engender performance-avoidance goals: The mediating role of autonomous motivation. *Journal of Educational Psychology, 103*(3), 683–700.

|| SIXTEEN

De-Grading Writing Instruction in a Time of High-Stakes Testing

The Power of Feedback in Workshop

P. L. Thomas

In junior high and high school, I was a full-fledged math and science nerd. I made A's in math and science classes; began to read, collect, and draw from Marvel comic books; and read voraciously science-fiction (SF) novels by Arthur C. Clarke as well as Larry Niven and Jerry Pournelle. My experiences in English class throughout junior high were mainly spent relentlessly plowing through grammar textbooks and diagraming sentences (all of which earned me A's) along with vocabulary workbooks and tests (which I ignored and as a result lowered my course grade to B's in English).

While I had life-changing experiences in sophomore and junior English because of my teacher, Lynn Harrill (Thomas, 2003), which I will deal with below, a moment from my senior year of high school English remains with me until this day. My teacher was Hilda Poole, a veteran and stereotypical English teacher—hair in a bun, humorless (at least in front of the students), and draconian about the rules of grammar and usage. I was generally known as a *good* student, graduating eighth in my class, and had a reasonably positive attitude about Mrs. Poole and her class.

As was common in the late 1970s (and to this day), the major project of any English class—and heightened in any student's senior year as graduation approached—was the research paper, an assignment fraught with pitfalls of failure and plagiarism. In fact, if my memory isn't overly exaggerating, the research project appeared *primarily* as a series of traps destined to cull the weak students from the graduation line. Two things about my experience with that senior research paper are unforgettable through the lens of my life and career—since I graduated high school intending to major in physics and was hired five years later as an English teacher in the high school from which I graduated.

First, my closest friends and I secretly workshopped and peer-edited our research papers. I note *secretly* because the acts were explicitly deemed cheating by our teachers—ironically in 1978–1979 just a few years after the National Writing Project (NWP) grew out of the Bay Area Writing Project begun in 1974 (History of NWP, 2012). Although my sophomore and junior English teacher, Mr. Harrill, would soon merge his professional career with the growing NWP movement, the view of writing as a *collaborative process* was absent during my secondary experience as a student.

Next, however, was the most significant event—the submission of my senior research paper. In the late 1970s, preparing a research paper to turn in for a grade was a Herculean act of typing skills, which I did not possess. Compounding and distorting the act of researching and then composing a documented essay for all of us at this time was the Modern Language Association (MLA) style sheet that called for footnotes (soon to evolve into endnotes, and then parenthetical citations). To prepare a paper, then, required typing a page, guessing at the space needed for a footnote, and retyping if the estimations failed.

This typing feat fell to my wonderful and patient mother, by the way.

And so I submitted my senior research paper, and days later, Mrs. Poole returned the essay, asking to speak with me. With tears brimming in her eyes, Mrs. Poole showed me the paper—I received an A/F—and then apologized for the inevitable F I received for grammar and mechanics since, due to a typing error (she acknowledged that the error was clearly a typing mistake), I had a fragment. During my high school years and well into the first years I taught at this same high school, the English department had a detailed and utterly bizarre scoring policy for essays. That scoring guide addressed only surface features—grammar, mechanics, and usage—and assigned point values for each "error."

The fragment received an automatic 31-point deduction, resulting in an F for the grammar and mechanics grade.

Mrs. Poole sat before me a veteran teacher, well respected and feared. She was clearly upset by the event but had abdicated her autonomy as a teacher to a mechanistic system for responding to student writing.

•

It is now 2012, and I am at the end of my first decade as a college professor of education. After 18 years teaching high school English, a career that was deep in my heart and bones as a *teacher of writing*, I moved to the university in part as an act of professional and scholarly autonomy. Teaching in education courses, however, has proven to be far less fulfilling and off-kilter to my central concerns with directly addressing human literacy—fostering writers.

After being allowed to teach one section of the university's introductory English course, I was fortunate that my university re-imagined its curriculum, replacing the two required first-year English courses with two first-year seminars designed to inspire and fuel student engagement in learning. One of the first-year seminars must be writing intensive, and the seminars are taught by professors across departments—not just the English faculty.

This curriculum change has afforded me a unique opportunity to teach a writing-intensive first-year seminar each fall at the university level, where I have the autonomy to implement writing workshop and, most significantly, to *de-grade the feedback process of my students crafting their essays*. In that context, this chapter opens with a brief discussion of how the writing curriculum has suffered a failed history in K–12 education—almost completely disconnected from the research and craft of composition as a field. Then, I detail my evolution as a teacher of writing from my high school years as a teacher and into my recent experiences with de-grading the writing classroom for first-year college students. I also examine how K–12 teachers of writing are both inhibited in best practices for composition because of the accountability era as well as how those teachers should and can reclaim the teaching of writing for all children.

Teaching Writing in the K–12 Classroom: "The Considerable Gap"

The field of composition is a fractured field, divided among elementary school writing (often conflated with teaching *handwriting*, but also teaching children to compose and conform to conventions), secondary school writing, college writing, journalism, and the traditional "creative writing" (see LaBrant, 1936, for a problematizing of the term). As Smagorinsky (2006) explains in his compilation of research on teaching composition: "The authors of the chapters in this book work in the thick of the *productive tensions provided by the competing schools of thought* [emphasis added]" (p. 12). Whether we consider the field of composition "fractured" or "in the thick of the productive tensions," "[c]omposition is . . . clearly the stepchild of the English department" (Kinneavy, 1971, qtd. in Smagorinsky, 2006, p. 1), revealing that teaching writing has historically struggled for *credibility* as a field and within the larger field of literacy or teaching English. Thus, while my discussion here is not intended to be exhaustive, I want to establish that the teaching of writing is far from being a monolithic concept, in part for somewhat bureaucratic and corrosive reasons and in part for valid reasons (a writer seeking to be a poet has a different agenda than a writer drawn to the field of journalism).

Fractured and marginalized, *teaching writing*, then, is defined by tensions and arguments about purposes and outcomes (see Hillocks, 2009, and Smagorinsky, 2009, for example); but school-based writing—especially from secondary school through undergraduate and graduate education—tends toward academic, scholarly, and expository modes of writing. In some ways, school-based writing is predisposed to functional modes of writing (as opposed to expressive modes, for example). But since this chapter seeks to focus on school-based writing, I want to emphasize what I believe to be the central aspects of the history of teaching writing in school—anchored by two claims from Lou LaBrant: "We have some hundreds of studies now which demonstrate that there is little correlation (whatever that may cover) between exercises in punctuation and sentence structure and the tendency to use the principles illustrated in independent writing" (1946, p. 127); and "A brief consideration will indicate reasons for the considerable gap between the research currently available and the utilization of that research in school programs and methods" (1947, p. 87).

While LaBrant as a scholar-educator focused primarily on reading (she advocated throughout her career free-reading programs), LaBrant's scholarship and practice reflect the historical grounding of the *disconnect* between the *research* in the field of composition and the *practices* of teaching writing in K–12 classrooms (LaBrant, 1936, 1946, 1947, 1950, 1953, 1955, 1957; Thomas, 2000, 2011a, 2011b). This historical "gap" between research and practice, as identified by LaBrant, represents K–12 writing instruction throughout the greater part of the 20th century until the 1970s. During LaBrant's many decades in education (1906–1971), most English classrooms focused on grammar textbooks and worksheets along with diagramming sentences, classifying isolated and direct instruction of language surface features as *writing instruction.*

The writing practices and curriculum of K–12 education ignored or rejected scholar-practitioners such as LaBrant until the rise of the National Writing Project in the 1970s and the influential impact of teacher-scholars that grew out of that movement, such as James Gray (2000, 2001), Donald Graves, Lucy Calkins, Ralph Fletcher, and Nancie Atwell (although these and other figures in this movement also represent the controversial nature and the tensions inherent in teaching writing in K–12 settings). Applebee and Langer (2009) identify a rise and peak of claims about implementing process writing, which we can associate along with the influence of the NWP as well as the National Council of Teachers of English (NCTE) as a *shift toward closing the gap between research and practice lamented by LaBrant*:

> Data over time also suggest that there has been some increase in emphasis on writing and the teaching of writing, both in English language arts classrooms and across the curriculum, although this may have begun to decline from its high. Process-oriented writing instruction has dominated teachers' reports at least since 1992, but what teachers mean by this and how it is implemented in their classrooms remains unclear. The consistent emphasis that emerges in teachers' reports may mask considerable variation in actual patterns of instruction. (p. 26)

Table 16.1. Writing Instruction Trends: Shifting Practices

1900–1970s	**1970s–1990s**
Writing instruction included, and was dominated by, direct and isolated grammar, mechanics, and usage instruction—often textbook-based; students completed few independent or extended compositions	Addressing surface features (grammar, mechanics, and usage) was embedded in and drawn from students' compositions; students composed more independent and extended original works
Prescriptive grammar, mechanics, and usage	Descriptive grammar, mechanics, and usage
Writing assignments driven by prompts chosen by teachers and designed to support students transmitting teacher-dispensed content back to the teacher	Writing assignments driven by student choice in both form and content
Writing modes dominated by text-analysis fulfilling paradigms of New Criticism and evaluated by how well content conformed to accepted claims	Writing modes drawn from authentic texts, and compositions created and evaluated by appropriateness for author purpose
Writer voice restricted to third-person—detached and academic	Writer voice expanded, including first-person and appropriateness for author purpose
Teacher's role was authoritarian and evaluative	Teacher's role was as a mentor and fellow writer
Students restricted to formal essays—thesis sentence; introduction, body, conclusion; five-paragraph essay	Students encouraged to draw writing forms from authentic texts as models; focus honored (instead of direct thesis), and paragraphing and organization chosen by appropriateness
Teacher-centered classroom instruction	Workshop format
Students worked in isolation; teacher feedback dominated by correcting errors and grades	Students drafting essays included peer-conferences and teacher-conferences (a writing community); teacher feedback focused on supporting student revision

Yet, as Applebee and Langer (2009, 2011) show, even this transformation of teaching writing during the 1970s, 1980s, and 1990s did not result in a unified practice or in an observable increase in students writing with purpose and autonomy in K–12 classrooms. However, there appears to be a relatively safe series of trends in how writing was assigned and taught throughout most of the 20th century, and then another series of trends associated with the NWP along with the rising influence of teacher-scholars. Table 16.1 is a tentative chart of those patterns, using evidence from several relatively recent examinations of teaching writing in the context of K–12 classrooms (Applebee & Langer, 2009, 2011; Brannon et al., 2008; Duxbury, 2008; Graham & Perin, 2007; Hillocks, 1995, 2002, 2003, 2009; Smagorinsky, 2006, 2009; Thomas, 2000, 2011a, 2011b, 2012; Wilson, 2010; Zemelman & Daniels, 1988; Zemelman, Daniels, & Hyde, 2005).

Again, this chart is not exhaustive or intended to be authoritative, but it shares many characteristics found in similar comparisons made by Weaver (1996) concerning the teaching of grammar in the context of writing and Zemelman, Daniels, and Hyde's (2005) characterization of how to transform a traditional writing classroom to a setting that honors research-based best practice. That said, broadly speaking, the teaching of writing at the K–12 levels was experiencing a significant shift and transformation into a rival content area to reading and math throughout the 1970s and into the 1980s. This field evolution remained contentious and continued to reflect LaBrant's (1947) charge that writing instruction pedagogy in practice remained too often removed from research in the effective teaching of writing (Applebee & Langer, 2009, 2011; Smagorinsky, 2009).

While I don't want to sound idealistic about an alternative development in the teaching of writing at the K–12 levels, I can state that whatever promise process writing, workshops, peer and teacher conferences, authentic texts, and descriptive attitudes toward language use held for the teaching of writing, that was all curtailed slowly but surely beginning in the overlapping decade of the 1980s—when the NWP was blossoming and *A Nation at Risk* under Ronald Reagan spurred the accountability era in education (Thomas, 2012b).

Accountability and the Death of Composition

"With the new federal education act, testing has become the official driving force to reform education," declared Hillocks (2003) about No Child Left Behind (NCLB), adding:

> This simple-minded view of education, with testing at its center, has been enacted into law. States that do not comply with the demand for testing will fail to receive federal money. Testing is here to stay for a long time, and my guess is that it is likely to increase. . . . When possible, the analyses need to examine the impact that testing has on teaching and the curriculum—that is, on how teachers and administrators decide to prepare students for testing. If enough people listen and are convinced by such analyses, perhaps it will be possible to improve testing practices, if not eliminate them. I am not sure I would bet on change, even in the face of a clear demonstration of the harmfulness of the tests. But I believe that the attempt is worth the effort. . . . The problem is that no matter how foolish the testing, it drives the writing curriculum and instruction in the state. In each state, we interviewed teachers in two large urban districts, as well as two suburban, one small town, and one rural district. The vast majority of these teachers, although complaining about the testing programs, tended to accept the tests uncritically. (pp. 63–64)

From 1983 until 2001, state-based accountability built on state standards and high-stakes testing spread like a cancer, enveloping, as Hillocks details, the curriculum and pedagogy of reading, math, and finally writing. Then, NCLB federalized the paradigm.

Throughout the 1980s, states began to incorporate writing assessments as part of the accountability mechanism, including requiring students to pass exit exams that often included tests of writing. This double-edged sword created a new primacy for writing—in other words, writing was authorized through its status as a test area along with reading and math—*and* corrupted what was just being *implemented* as best practice in writing. As an example, but a representative one, in my home state of South Carolina throughout the 1980s, students had to pass exit exams in high school, first administered in 10th grade and assessing reading, math, and writing. The writing exit exam was a writing sample, had no time limit, and encouraged students to write multiple drafts, included in the submitted packet of work.

As Hillocks (2003) uncovered, however, the potentially positive impact of the state assessments of writing—recognition of the value of composition across disciplines, honoring writing samples over multiple-choice tests, fostering writing as a drafting process—was soon washed away as writing instruction for all students in grades leading up to 10th grade and for students who failed the first attempt of the exit exam quickly devolved into prescribed, prompt-driven writing (prompts derived from the state departments of education) scripted by the state rubric for scoring as a template. As soon as the mid-1980s, the test-prep assault on writing instruction had eroded what was just beginning to blossom as best practice in process, workshop-based composition in K–12 classrooms.

Because of the high-stakes testing paradigm, writing instruction and products by students conformed to dynamics that nearly universally contrasted with authentic composition:

- Generic prompts that allowed students to show content proficiency in standards-based modes of writing (exposition, persuasion, narration, and description, for example)
- Rubric-driven assessment of writing that specifically delineated every aspect of writing and re-emphasized the weight of conforming to prescriptive views of Standard English
- Mechanistic patterns of essay development reflecting the broad elements of the traditional five-paragraph essay (although in South Carolina, students were encouraged to write a mere three paragraphs as to lessen the opportunity to make mistakes)

Writing instruction fell into lockstep with other fields under the weight of high-stakes accountability: Classroom practice was dictated by the holy grail of raising test scores, and as a result, best practice be damned. Hillocks (2003) revealed that in Illinois, writing instruction became bound to the state standards and the definitions inherent in those standards, the theory of writing mandated by those standards (or the *failure* to identify a theory), the quality (validity) of the prompts and tests themselves, the scoring rubrics and process, and the benchmark essays (replacing authentic models) used to guide the instruction and scoring. In Illinois, Hillocks noted that the five-paragraph essay dominated instruction, a regrettable consequence of the accountability era:

> This is a crucial time in American democracy. We are faced with problems that demand critical thinking of all citizens. We need to help students examine specious arguments and know them for what they are. *Our tests encourage the opposite* [emphasis added]. They encourage blurry thinking and obfuscation. As a society, we cannot afford to spend valuable classroom time on vacuous thinking and writing. We need to tell citizens and legislators what these problems are and insist that they be addressed. (p. 70)

While the impact of state-based high-stakes testing of writing was powerful, the first two decades of accountability were soon to be eclipsed in the next decade—again by testing in the form of National Assessment of Educational Progress (NAEP) and the addition of a timed writing section on the SAT in 2005. As state assessments came under additional scrutiny throughout the accountability era begun in 1983 (Finn, Petrilli, & Julian, 2006), the 2000s saw a rise in national benchmarks for measuring the quality of state accountability systems; thus, NAEP grew in importance, but as I examined (Thomas, 2004a, 2004b) the writing prompts, sample graded essays, and scoring rubrics, I found that NAEP failed basic standards of assessment as well as exhibiting Hillocks's (2003) criticisms of the exact state tests NAEP was used to validate. The 2002 NAEP data on writing—which I later correlated with the 2007 test—revealed conflating of assessing writing proficiency with reading proficiency (the prompts compromising the *validity* of the assessment), cultural bias in the prompts, and jumbled messages in the rubrics and scoring guides when the terminology was examined against the scoring directions and then compared to the sample scored student essays.

But the problems with NAEP writing assessments to judge the quality of state assessments of writing were less dramatic than the College Board's addition of writing to the SAT, a move preceded with a great deal of fanfare and public relations (National Commission on Writing in America's Schools and Colleges, 2003). The College Board first reported that the United States had a national failure to teach writing, and then it conveniently added writing to the SAT to address the claimed weaknesses in writing instruction. *The media and the public accepted this move uncritically, failing to identify that if writing instruction was floundering in the United States as the College Board claimed, the primary blame for that failure was already identified by Hillocks (2002, 2003): high-stakes testing.* What logic, then, justified the College Board's offering additional testing of writing to address the deterioration of writing instruction as the result of state-level high-stakes testing?

As well, the National Council of Teachers of English (NCTE) challenged the use of timed tests (Ball et al., 2005) and their impact on writing instruction and student compositions. The report identified four areas:

1. Concerns about the validity and reliability of the test as an indication of writing ability
2. Concerns about the impact of the test on curriculum and classroom instruction as well as on attitudes about writing and writing instruction
3. Concerns about the unintended consequences of the uses of writing tests
4. Concerns about equity and diversity (p. 2)

The SAT writing test added new problems, as well, to teaching writing since the test sent a corrosive message with its format: a 25-minute, one-draft, prompted essay and a section with 35 multiple-choice questions. This new writing test was more than half selected-response, and that alone renewed the traditional focus on teaching and assessing writing surface features in isolation. Further, Newkirk (2005) has exposed that the SAT writing sample rubric has led to student coaching to fabricate evidence since the rubric requires the *use* of evidence regardless of its validity or accuracy. The writing section of the SAT also increased the move toward computer-scored writing, which reduces all feedback on writing to a grade:

> How can a computer determine accuracy, originality, valuable elaboration, empty language, language maturity, and a long list of similar qualities that are central to assessing writing? Computers can't. We must ensure that the human element remains the dominant factor in the assessing of student writing. (Thomas, 2005b, p. 29)

Wilson (2006a) challenged computer-scored writing from the Educational Testing Service to expose that computer scoring of writing fails both authentic writing forms and valid evaluations of the qualities that authentic writing forms exhibit. Wilson first placed her own experiences as a writing teacher within the context of why the allure of computer scoring of writing existed:

> Five years later, I still haven't given students in my writing courses a multiple-choice final exam. I'd like to chalk up my resistance to idealism and strength of character, but it's partly an issue of logistics. I've got to keep the students occupied for an entire hour and a half. How many multiple-choice questions can you devise about the steps of the writing process? Secretly, though, I've held out

> hope; if technological progress can send a probe to Mars and pack 15,000 of my favorite songs into an iPod, why can't it devise a way to grade student writing?

After explaining how computer scoring of essays works—computers essentially use the characteristics of scored essays to drive the scoring of new essays; thus, a great deal depends on the essays being scored conforming to the paradigms imposed by the computer program—Wilson noted that ETS claims computer-scored essays are more accurate than human grading. (And this claim has been repeated since, although few ever explain that this correlation is judging how well the computer scores against a rubric compared with a human. In other words, computers are more consistent at applying a rubric than humans, but that doesn't mean that the evaluations are accurate against the fuzzy but real rubric of "authentic writing.")

The genius of Wilson's (2006a) skepticism about computer-scored writing was her submitting Sandra Cisneros's "My Name" from *House on Mango Street.* The result?

> Criterion delivered perfectly on its first promise; it created a printout of responses about grammar, mechanics, style, organization, and development in less than thirty seconds. But Criterion was not overjoyed by Cisneros' writing; the first thing that struck me about the feedback was that it offered no praise. Praise is important in my own feedback because it helps students become better writers. When I sit down to write, about 90 percent of what spills out is not suitable for public consumption. Part of becoming a good writer is the ability to recognize the 10 percent with potential and the courage to purge the rest. When I praise part of a student's paper, I am pointing out the standard they have set for themselves. But Criterion cannot recognize what writers do well. It only defines and describes deficits, and its feedback to Cisneros merely pointed out "mistakes" in "My Name"—repetitious word use, use of fragments, and problems with organization and development. What would happen to "My Name" if Criterion's immediate feedback had shaped it?

Then, Wilson revised Cisneros's piece, resulting in a discernably worse passage (notably longer and shaped into the five-paragraph format), but a piece of writing that conformed to the rubric. Wilson also identified how complex and powerful the role of testing and computer-based scoring of writing proves to be—even until 2012:

> If the considerable time and effort that has been put into computer grading systems that *produce standardized writers* [emphasis added] had been put into reducing instructors' loads, perhaps we wouldn't be so desperately looking to testing corporations for solutions to problems that they have created. But the *corporations that profit from the perception of our incompetence* [emphasis added] can't let that happen. *If they trusted teachers to teach and if they trusted students to think and question, they'd be out of a job* [emphasis added].

And this leads to the next important development that killed the growth of teaching writing: the rubric. The accountability era either created or helped support the parallel rise of "backward design" approaches to curriculum and instruction that argued for starting most, if not all, educational decisions with the assessment (Wiggins & McTighe, 1998). This backward design approach was couched in part as a response to the *gotcha* element of traditional teaching and assessment in which teachers withheld the details of grading and testing to compel students to study everything to ensure that they could answer what happened to appear on the test. For writing instruction, anchor essays, sample graded essays, and ultimately rubrics became the driving force behind what students wrote, how they wrote the essays, and how those essays were graded (along with what teachers offered as feedback).

Wilson (2006b, 2007, 2007/2008) and Kohn (2006), like Hillocks's work on standardized testing of writing, confronted the corrosive impact of rubrics on writing instruction. Wilson (2006b) noted:

> I had come to see specific feedback as one of the most important ways I could help students to become better writers. After all, the purpose of writing is to create a response in the reader's mind. After I set up meaningful writing opportunities for students, I owed them the chance to see what happened in my mind as I read their words. (p. 63)

But rubrics replaced rich and specific feedback with the pursuit of efficiency and standardization. For Wilson (2006b), however, honoring authentic theories of texts, composing, and teaching writing must replace the pursuit of efficiency:

> It is startling that focusing on our response to the writing and thinking about how we can help the writer improve is, in fact, going against the grain. But writing itself has always presented a problem for positivist testing specialists; it has always been too messy and subjective to satisfy their need for factory-style assessment (Lynne). Rubrics are writing assessment's current sacred cow because they provide the *appearance of objectivity and standardization that allows direct writing assessment a place in standardized testing programs* [emphasis added] (Broad). By accepting the standardized responses inherent in rubrics, we undermine the power of the experiences of reading and writing. In the end, assessment must be a conversation—just as writing exists for the purpose of conversation. I'm not willing to let rubrics script that conversation for me. (p. 66)

If we consider the chart detailed above showing the transition from traditional approaches to teaching writing through the 1970s and then the rise of best-practice, process writing in the 1970s–1990s, the accountability era turned that around and *increased* the technocratic elements of the traditional paradigm. The most powerful and corrosive elements of the accountability era—the rise of state and national tests of writing—include the following: prompt-driven compositions with content and mode determined for the students, rubric-driven feedback and assessment of writing, anchor essays supplanting authentic texts, single-draft essay samples to assess writing proficiency, and the return of isolated, selected-response assessments honored as evaluations of writing proficiency.

My career as a teacher of writing, then, overlaps that accountability era, and my journey to de-grading the writing classroom has been strongly influenced by and reacting against these paradigms, as I now detail below.

The (R)Evolution: From Degrading Students to De-Grading the Writing Process

In the fall of 1984, I walked into the classroom in which I was a student of Lynn Harrill's from 1976 until 1978. By 1984, South Carolina led the United States with a few other states on the front edge of the accountability era spurred by *A Nation at Risk* (Thomas, 2012b), implementing the Basic Skills Assessment Program (BSAP) that included elementary, middle school, and high school tests of math and English with the high school assessments as highest-stakes exit exams. As I mentioned above, the writing assessments were some of the first in the country as part of this era of accountability, and in the school where I taught, the writing section of the exit exam became the default writing curriculum—unless a teacher purposefully worked against that momentum.

Our small, rural school with high poverty achieved several years of 100% passing rates for the exit exams, ranking first in the state multiple times, so our test-prep culture was reinforced by those circular accolades (including a banner each year proudly displayed in our lobby). Compounding this school culture was my initial and uncritical affinity for mastery learning, discovered and embraced during the master's work I was completing in my initial year of teaching. In the first five or six years of my teaching, I was unable to recognize the contradiction between my heart (student-centered, emancipatory) and my deeply analytic brain, trained by my success as a compliant student.

In the foundational decade of teaching, I was from the first day primarily a teacher of writing, couched inside my role as an English teacher. My students always wrote large amounts and wrote a variety (far too much in the beginning, in fact). And I was perpetually experimenting and revising to make myself a better teacher of writing; when student essays disappointed me, I always assumed I was failing my students. There existed a few cracks in my early teacher persona (authoritarian), including my recognition that grammar books and isolated instruction of surface features were not teaching writing, that isolated vocabulary instruction was not improving students' literacy, and that school-only modes of writing (such as the five-paragraph essay) were not the goals of an authentic classroom.

My first attempts to improve my feedback to student writing and to increase the quality of students' drafting and final products included building a framework of elements I commonly noted and responded to in student essays; in fact, I jotted notes on a 10 by 13 envelope used to bundle each set of essays. From that authentic practice grounded in student writing, I created a list of about 50 numbered items (1 was commas, 19 was vague word usage, 41 was discussion not connected to a thesis/focus, for example) that I correlated with pages in the students' grammar textbook. I felt this was efficient (handwriting the numbers beside highlighted or underlined elements of the student's essay) and authentically transformed the grammar text into a resource.

But I was wrong.

I saw little change in how well students revised; I also saw no real gains in first submissions over the course of the year. What did I discover in this first phase? Students admitted they used the one-page list of numbers with revision and editing needs identified, but they also confessed they never opened the grammar text. *Lesson One: An inauthentic resource (prescriptive grammar texts) cannot be transformed into an authentic one.* Also, I learned much later that student outcomes must be viewed in the context of the teacher's assignment. *Lesson Two: Student writing samples often reflect what teachers prescribe, assign, and honor, but not necessarily student competencies or proficiencies.*

Student first drafts and revisions, then, were much greater reflections of my teaching than my students' learning. Thus, I moved into phase two of my evolution as a teacher of writing. Phase two included a few key characteristics and changes in pedagogy. First, this transformation was greatly boosted by my participation in the Spartanburg Writing Project (SWP) in the late 1980s, notably under the guidance of Brenda Davenport. More than a decade later, Brenda and I would be co-lead instructors for the SWP, but when I was a workshop participant, Brenda asserted her best-practice knowledge base and rattled my teacher-centered and authoritarian attitudes toward language and teaching writing. In short, she put me in my place, and after my initial hesitancy, my perspective was changed dramatically.

Key to my transition was being not just a *writer who was a teacher of writing* (which I thought, arrogantly, was what all teachers should be) but a *scholar of teaching writing who writes and teaches*. This is a central principle of the NWP and one that has served me well, and would serve the education reform movement well if that movement were not committed to de-professionalizing teachers. Part of that transformation included my second effort at the grammar book dilemma. First, I ditched the traditional grammar book (we used the classic Warriner text); students were instructed to use that text to line the bottom of their lockers (several years later, as department chair, I was successful in stopping grammar textbooks being issued to all students, replacing that with providing each teacher a classroom set to keep if they wished to use them; we did the same with vocabulary textbooks). I used department funds, then, to purchase PermaBound volumes of William Strunk Jr. and E. B. White's *The Elements of Style*.

Next, I embraced my new perspective as a scholar-teacher by drafting and implementing a writing textbook with and for my students. I took the number system I had correlated with the grammar text and revised it by writing a writing text opening section and then expanding the numbers to include non-technical explanations of each number (such as commas or vague word choice) with examples of sentences and passages needing revision (taken from student essays) and sentences and passages revised. (This book was photocopied and distributed to all my students, and has since then been published [Thomas, 2001] and become a blog: http://conventionallanguage.blogspot.com/.)

During my second phase as a teacher of writing, I evolved toward an awareness of the importance of authentic models of writing, contextualized but direct instruction, student choice and autonomy, and detailed but supportive feedback that encouraged revision. My self-made textbook, however, did not work as well as I planned (and neither did the prescriptive nature of *Elements of Style*) since students also admitted to not looking at the explanations and examples any more closely than they had the correlations with the traditional grammar text. But I did see improvements—although likely from the sheer volume of reading and writing my students experienced. Important in this phase is also my letting go of New Criticism (Thomas, 2012a) as the foundational approach to text and expanding for my students a wide array of lenses through which to make meaning from text.

I continued to struggle with the evidence of my practice—the quality of the first submission of each essay and the quality of each revision. A significant part of the experiment during both the first and second phases of my evolution as a teacher of writing was how I assigned and calculated grades. I offered articulated grades on essays that addressed both the quality of broad characteristics of their writing as well as made clear what aspects of writing were of greatest value. The grading system I used most of my career as a high school writing teacher was a 20-point scale, assigned as follows: Content/organization = 10 points, Diction/style = 5 points, and Grammar/mechanics = 5 points. This scale showed students that three-fourths of the value of their writing was what they said, how they organized their content, and how they expressed those messages in the words, sentences, and paragraphs they formed—with surface features far less significant by their weighting.

Another aspect of assigning grades was the tradition of averaging grades among all assignments, which I first confronted and then rejected (Wormeli, 2006). That evolution included several experiments. At first, I assigned an articulated grade to each submitted draft of student essays and implemented a graduated approach to determining grades that honored the last grade on each essay, all of which were then averaged, in the first quarter, and then increasing how many drafts counted in the averaging as the year progressed by increasing their accountability through moving grades earlier and earlier in the drafting process. This technique appeared to me to increase student accountability for their work, but I failed to consider the relationship between increased accountability and the need to experiment freely in the discovery aspect of drafting. Once I came to terms with the manipulation in this process (see below) in terms of external rewards/punishments and their corrosive influences on learning (Kohn, 2011), I noticed that this technique made some statistical sense but almost no pedagogical sense.

Later, I stopped assigning grades to the first and even subsequent drafts of student essays, requiring students to identify when they submitted a "last" draft; then I would assign a grade. I still tended to average these terminal grades for each essay. I did find that students appeared to work more carefully on their revisions once I shifted their focus away from "what is my grade" and toward "how can I revise this essay to be stronger." The structures of public school, however, never allowed me to completely de-grade the writing classroom.

The first two-thirds of my evolution as a teacher of writing included important aspects of coming to know and understand not just being a writer but being a *teacher* of writing. All of that included the powerful element of being a practitioner-scholar, which I drew from the SWP/NWP, but the key transitional point for me was my doctoral program in the mid-1990s. My doctoral program allowed me to discover and implement critical pedagogy and critical constructivism as underpinning philosophies and theories to drive the continued evolution of my teaching of writing. As well, my doctoral program introduced me to Joseph Williams's *Style*, which supported my transition away from prescriptive grammar to descriptive grammar. *I came to frame all of my writing pedagogy and assessments within my agency and autonomy as a teacher as that impacted and overlapped with the agency and autonomy of my students*. Authentic and original writing were both the goals of my curriculum and an avenue to larger goals addressing democracy and empowerment.

Central, then, to the larger (r)evolution of me as a teacher of writing centered around *grades*—and thus the extended path I have taken to examine here at the end the de-graded writing classroom at the university level.

I received my doctorate in 1998 and remained in the high school English classroom until 2002, when I moved to teacher education. That move afforded me much greater autonomy in terms of my pedagogy and my assessments and course grades. As a high school teacher, I had eventually abandoned averaging grades and had stopped entirely giving any selected response or traditional tests. But grades and how those grades were calculated and assigned were far more prescribed and monitored than at the university level (although I must caution that my higher education experience includes different grading issues such as an obsession by the administration with grade inflation).

Along with the pedagogical and assessment autonomy I experience as a professor (now tenured), the university's transition to first-year seminars has influenced greatly my practices and offered ample evidence about how de-grading a writing classroom works, and doesn't. In all of my university courses, in fact, I refuse to put grades on assignments throughout the semester. Instead, I have two practices: (1) I provided ample feedback and require and allow students to revise most assignments until the students are pleased with the work, and (2) I invite and urge students to arrange conferences as often as desired throughout the semester to discuss their grades (what their grades would be if assigned, what they would assign themselves, and what I anticipate they will be assigned at the end of the course).

De-grading the writing class and encouraging a conversation about grades instead of labeling assignments with grades have combined to lift the effectiveness of my writing instruction significantly because these practices reinforce the autonomy and agency of the students and shift the focus of the classes to the quality of the compositions and the growth of the students as writers and away from courses as credentialing. Yet, several problems still remain in the de-graded writing classroom, and they include the following:

- Students remain uncritical of their behavior as *students* as opposed to *learners* or *humans*. Student behavior is powerfully couched in behaviorism and capitalism paradigms that are dehumanizing and inauthentic. Thus, to de-grade the writing classroom is to confront and reject the wider culture of being a compliant student as a gateway to being a compliant worker. Conversations with students about these tensions help address these hurdles, but I still struggle against them.
- Even more autonomous environments (higher education versus K–12 settings) remain tethered to grades; thus, a de-graded writing classroom is only *temporarily* de-graded. Ultimately, grades must be calculated, assigned, and even justified. That said, de-graded writing classrooms can offer evidence for challenging institutional commitments to grades since the de-graded classroom can showcase high-quality student writing instead of reductive grades (see a recent course blog of students' original work: http://thomasfywfurman.blogspot.com/).
- Nontraditional practices in any classroom make direct and indirect commentaries on other classrooms, the practices in those classrooms, and the teachers/professors leading those classrooms. Again, that tension should not be a justification for grading, but anyone de-grading a writing class

should have this conversation with students also—an extended conversation about the norm of grading and what that reinforces and honors.

- Many students are unconsciously finishers and see all academic behaviors as assignments; thus, students are apt not to appreciate or fully commit to the writing process, including a failure to draft often and over extended periods of time, hesitating to share draft work with peers and the teacher/professor, and persisting in viewing the products of their learning (such as essays) as means to a grade and not goals themselves.

Yet, despite these hurdles, I have discovered that authentic writing instruction depends on re-imagining feedback on student drafts and ultimately de-grading the workshop environment. Let me end, then, with some guiding principles for de-grading the writing classroom:

- Reductive responses to student writing (grades, rubric scores) fail to enrich the writing process (see Gould, 1996, regarding singular quantification of complex processes). *Teacher feedback must be rich, detailed, and targeted to support revision.* The most powerful feedback includes identifying key strengths in a student's work ("Do this more often!") and questions that help guide students toward revision ("Why are you omitting the actual names of your family members in your personal narrative?"). To share with students the specific and contextualized characteristics that constitute an evaluation (such as an A) provides students as writers the evidence needed to build rubrics of expertise for future writing, and learning.
- Teacher and student roles in the de-graded writing classroom must be revised—*teachers as authoritative, not authoritarian; teacher as teacher/student; and student as student/teacher* (Freire, 1993). The de-graded writing classroom allows a balancing of power that honors the teacher's agency as a master writer and master teacher of writing without reducing the status of the student as beneath her or his agency and autonomy, both of which are necessary for the growth of any writer.
- *De-grading the writing classroom increases the importance and impact of peer conferencing by removing from the teacher as the primary or pervasive role of evaluator.* Feedback from peers and from the teacher become options for students as they more fully embrace their roles as process writers.

•

This chapter rests inside a volume dedicated to challenging the degrading and detesting aspects of tests and grades in education by calling for acts of de-grading and de-testing our schools. Here, I have couched the pursuit of a de-graded writing classroom within the context of the history of teaching writing as well as the current three-decades-long accountability era.

My experiences as a teacher of writing have informed my argument that grades and tests themselves are corrosive to the pursuit of education among free people dedicated to the values of democracy, agency, and autonomy. Some remain committed to the belief that it isn't grades or tests but how they are used.

I reject that compromise.

Tests and grades are by their nature *reductive*, in that they necessarily reduce complex and nuanced human behaviors into one metric. Early in my teaching career, I recognized that three essays labeled as B- work were as different from each other as three essays labeled A, B, and C. Inherent in the failure of grades and tests is the act of averaging, which washed away the nuance of multiple data points into, again, one data point. Consider my grading scale I used for years:

Essay 1
Content/organization: 5
Diction/style: 4
Grammar/mechanics: 4
Total: 13
Essay 2
Content/organization: 9
Diction/style: 3
Grammar/mechanics: 1
Total: 13

These two essays have the same singular grade, but the quality of each is significantly different, and the revision needed is also distinct.

De-grading the classroom, then, is about seeking authentic teaching and learning, but it is also about power. Grades and tests have shifted our teaching and learning gaze away from teachers and students, and in effect shifted power to data, instead of people. It is with that argument that I end, calling for the de-graded classroom as an act of honoring the human element of teaching and learning while also rejecting the failure of grades and tests in that process.

References

Applebee, A. N., & Langer, J. A. (2009). What is happening in the teaching of writing? *English Journal, 98*(5), 18–28.

Applebee, A. N., & Langer, J. A. (2011). A snapshot of writing instruction in middle schools and high schools. *English Journal, 100*(6), 14–27.

Ball, A., Christensen, L., Fleischer, C., Haswell, R., Ketter, J., Yageldski, R., & Yancey, K. (2005, April 16). *The impact of the SAT and ACT timed writing tests*. Urbana, IL: NCTE.

Brannon, L., et al. (2008). The five-paragraph essay and the deficit model of education. *English Journal, 98*(2), 16–21.

Duxbury, A. R. (2008). The tyranny of the thesis statement. *English Journal, 97*(4), 16–18.

Finn, C., Petrilli, M. J., & Julian, L. (2006). The state of state standards. Thomas B. Fordham Institute. Retrieved from http://www.edexcellencemedia.net/publications/2006/200608_thestateofstatestandards2006/State%20of%20State%20Standards2006FINAL.pdf

Freire, P. (1993). *Pedagogy of the oppressed.* New York: Continuum.

Gould, S. J. (1996). *The mismeasure of man.* New York: Norton.

Graham, S., & Perin, D. (2007). *Writing next: Effective strategies to improve writing of adolescents in middle and high schools—A report to Carnegie Corporation of New York.* Washington, DC: Alliance for Excellent Education.

Gray, J. (2000). Teachers at the center: A memoir of the early years of the National Writing Project. *The Quarterly, 22*(4). Retrieved from http://www.nwp.org/cs/public/print/resource/41

Gray, J. (2001). James Gray on coaching: An excerpt from *Teachers at the center. The Voice, 6*(3). Retrieved from http://www.nwp.org/cs/public/print/resource/167

Hillocks, G., Jr. (1995). *Teaching writing as reflective practice.* New York: Teachers College Press.

Hillocks, G., Jr. (2002). *The testing trap: How state writing assessments control learning.* New York: Teachers College Press.

Hillocks, G., Jr. (2003). Fighting back: Assessing the assessments. *English Journal, 92*(4), 63–70.

Hillocks, G. (2009). Some practices and approaches are clearly better than others and we had better not ignore the differences. *English Journal, 98*(6), 23–29.

History of NWP. (2012). National Writing Project. Retrieved from http://www.nwp.org/cs/public/print/doc/about/history.csp

Kohn, A. (2006). The trouble with rubrics. *English Journal, 95*(4), 12–15.

Kohn, A. (2011, November). The case against grades. *Educational Leadership.* Retrieved from http://www.alfiekohn.org/teaching/tcag.htm

LaBrant, L. (1936). The psychological basis for creative writing. *English Journal, 25*(4), 292–301.

LaBrant, L. (1946). Teaching high-school students to write. *English Journal, 35*(3), 123–128.

LaBrant, L. (1947). Research in language. *Elementary English, 24*(1), 86–94.

LaBrant, L. (1950). The individual and his writing. *English Journal, 39*(4), 185–189.

LaBrant, L. (1953). Writing is learned by writing. *Elementary English, 30*(7), 417–420.

LaBrant, L. (1955). Inducing students to write. *English Journal, 44*(2), 70–74, 116.

LaBrant, L. (1957). Writing is more than structure. *English Journal, 46*(5), 252–256, 293.

National Commission on Writing in America's Schools and Colleges. (2003). *The neglected "R": The need for a writing revolution.* New York: College Board. Retrieved from www.writingcommission.org

Newkirk, T. (2005, November). The new writing assessments: Where are they leading us? *English Journal, 95*(2), 21–22.

Smagorinsky, P. (Ed.). (2006). *Research on composition.* New York: Teachers College.

Smagorinsky, P. (2009). Is it time to abandon the idea of "best practices" in the teaching of English? *English Journal, 98*(6), 15–22.

Thomas, P. L. (2000, January). Blueprints or houses?—Looking back at Lou LaBrant and the writing debate. *English Journal, 89*(3), 85–89.

Thomas, P. L. (2001). *Vivid language: Writer as reader, reader as writer.* Lanham, MD: University Press of America.

Thomas, P. L. (2003, November). A call to action. *English Journal, 93*(2), 67–69.

Thomas, P. L. (2004a). *Numbers games—Measuring and mandating American education.* New York: Peter Lang.

Thomas, P. L. (2004b, October). The negative impact of testing writing skills. *Educational Leadership, 62*(2), 76–79.

Thomas, P. L. (2005a). *Teaching writing primer.* New York: Peter Lang.

Thomas, P. L. (2005b, May). Grading student writing: High-stakes testing, computers, and the human touch. *English Journal, 94*(5), 28–30.

Thomas, P. L. (2011a, January). "A respect for the past, a knowledge of the present, and a concern for the future": The role of history in English education. *English Education, 43*(2), 123–144.

Thomas, P. L. (2011b, September). Revisiting LaBrant's "Writing is more than structure" (*English Journal*, May 1957). *English Journal, 101*(1), 103–104.

Thomas, P. L. (2012a, January). "A richer, not a narrower, aesthetic"—The rise of New Criticism in *English Journal. English Journal, 101*(3), 52–57.

Thomas, P. L. (2012b, May 2). Misreading the achievement gap: A tale of bi-partisan failure. *AlterNet.* Retrieved from http://www.alternet.org/education/155233/misreading_the_achievement_gap%3A_a_tale_of_bi-partisan_failure

Weaver, C. (1996). *Teaching grammar in context.* Portsmouth, NH: Heinemann.

Wiggins, G., & McTighe, J. (1998). *Understanding by design.* Alexandria, VA: ASCD.

Wilson, M. (2006a, Spring). Apologies to Sandra Cisneros. *Rethinking Schools Online, 20*(3). Retrieved from http://www.rethinkingschools.org/archive/20_03/apol203.shtml

Wilson, M. (2006b). *Rethinking rubrics in writing assessment.* Portsmouth, NH: Heinemann.

Wilson, M. (2007). Why I won't be using rubrics to respond to students' writing. *English Journal, 96*(4), 62–66.

Wilson, M. (2007/2008, December/January). The view from somewhere. *Educational Leadership, 65*(4), 76–80.

Wilson, M. (2010). Rethinking a writing teacher's expertise: Following students under the kitchen table. *English Journal, 99*(3), 50–56.

Wormeli, R. (2006). *Fair isn't always equal: Assessing and grading in the differentiated classroom.* Portland, ME: Stenhouse.

Zemelman, S., & Daniels, H. (1988). *A community of writers: Teaching writing in the junior and senior high school.* Portsmouth, NH: Heinemann.

Zemelman, S., Daniels, H., & Hyde, A. (2005). *Best practice: Today's standards for teaching and learning in America's schools* (3rd ed.). Portsmouth, NH: Heinemann.

SEVENTEEN

One Week, Many Thoughts

Brian Rhode

Have you ever had the pleasure of watching a school bloom? I have. I watched the walls around me burst into color, like flower petals extending themselves to the great warmth of the spring sun. Splashes of primary shades crawled throughout the school thoroughfares in which I spend my days as a professional. The entrances to classrooms became bustling hives of activity, and the productivity was evidenced in the variety of posters, pictures, and projects that emerged. Suddenly my small elementary school in upstate New York resembled a field of flowers in the full throes of its spring awakening!

I am certain many of you are asking what ignited such a school wide explosion of creativity. Quite simply, it was the result of a week without testing. My principal, Dr. Peter DeWitt, had the idea back in the fall of 2011 to give us, as a staff, a much-needed break from the relentless drive of standardized assessment-based instruction. As a veteran of the classroom himself, he recognized a way to reinvigorate his teachers by endorsing a respite from the type of instruction that seems to stand in a starkly antagonistic position to the attitudes and beliefs that typically bring people into teaching.

Needless to say, the No-Testing Week became a huge success. Interestingly, for me it also became a catalyst for reflecting on the environments and relationships that are created by a school climate that is or is not based on high-stakes assessment.

One might assume from my opening image that the current culture feels to me like a cold and harsh winter without apparent end. Frankly, I am okay with that. What I will argue for in this essay is that there might be more far-reaching intellectual and inter-relational damage among teacher-to-teacher, teacher-to-student, and student-to-learning interactions at the hands of high-stakes testing. This assessment culture is not benign, and its affects on education could push us into the type of stark learning-winter that we may never emerge from.

A Glimmer of Hope

In the fall of 2011, as I mentioned above, I had the pleasure of participating in a week without testing, an idea put forth by Peter DeWitt, my principal. We work together at Poestenkill Elementary School in Poestenkill, New York. The concept was simple—spend a week instructing without the end goal being a test or the collection of some measurable data. He encouraged us to be creative and enjoy a week without worrying about prepping students for paper/pencil assessments or standardized tests. He primed us all with motivational speeches recalling the days when we felt the freedom to be creative with our instruction and teach with engaging projects. Eventually, enough people realized this was not some stiff mandate; rather, it was an invitation to engage students and motivate them for a week through creativity, and many began to rally behind his idea.

In my own practice, which has spanned more than 10 years of teaching and involvement in education, I strive to connect my instruction to students' lives beyond those skills they need to test well and meet prescribed skill or content checkpoints. I have noticed an increasing irony between how students are actually tested and the skills that are being emphasized as part of a strong 21st-century curriculum. For instance, a significant theme among new curriculum design is collaborative problem solving. This is sensible; my team and I work together closely to collectively improve our craft and solve issues as they arise. Yet, there is absolutely no part of the current standardized tests used for ELA and math in New York State that has a collaborative component. I am certain that many would be happy to explain to me why group test-taking is not necessarily a good idea for finding out a student's potential skill set, but the discrepancy for me is still glaring. Suffice it to say, I was excited about the week without testing.

In no time at all our suburban elementary school became an eclectic collection of projects, illustrations, models, and many other items. In classrooms and throughout the hallways was the evidence of enthusiastic inquiry into various topics ranging from author's studies of particular books to Social Studies projects about Native Americans. Our cyber-hallways also became filled with projects; my fifth graders completed computer-based projects as part of a culminating set of activities at the close of a unit about government. Teachers, students, and visitors marveled at the showcase for creativity our school had become, and the seeds of a real sense of accomplishment began to grow.

About midway through the week I noticed a significant change in the whole aura of our school. Thus far the year had been quite stressful, for various reasons.

Within the fifth grades throughout our district, class sizes approached 30 students per class, due mostly in part to dramatic cuts in spending to school districts by the New York State Education Department. Trying to manage our professional duties with these increased caseloads became a regular topic of conversations among the fifth-grade team both in my school and across the district. I cannot speak for everyone in my school, but my impression was that our whole community experienced more stress on a regular basis. This backdrop helped the changes in attitude brought about by our week without testing stand out.

Simply put, my colleagues seemed happier, as did I. Regularly I hear those I work with express how tired this job can make them, and I often agree. However, during this test-free week the same sentiment was being expressed with smiles and tones that were undergirded by a sense of positive accomplishment, which was so refreshing and energizing.

By Thursday of that week this shift had become so obvious to me that I wanted to share my thoughts with Dr. DeWitt. I went into Peter's office and sat down to chat, as I often do, and I burst into an excited retelling of my observations about the event and how invigorating it seemed to be for everyone. As it turned out, he was feeling and seeing the same change. We spoke excitedly of how great the week had turned out and marveled at the power such a simple shift in philosophy could bring.

Despite all of this excitement, the week was a short-lived respite, and as with all good things, it had to come to an end. Still, an experience like that never goes away completely. Since that week I have returned many times to the feelings and observations that came from participating. I wanted to understand what lay at the core of the difference between the energetic environment of a week without testing and the typical climate of test-based schooling that seems to drain enthusiasm. My reflection led toward differences that are, I believe, stark and caustic toward a truly positive school culture.

The Exaggeration of Antagonism

Life in classrooms is becoming a bristly environment of shaming and blaming. Success is everyone's goal, but failure is treated like a hot potato; no one wants to admit responsibility. My professional and personal observations are showing me that the partnerships between home and school and even between students and teachers are becoming more strained.

High-stakes testing, especially those linked to teacher and school evaluations, are expanding a resistance culture. This dialing up of tension is washing out the hope of having truly positive academic classrooms. Students and teachers no longer have the luxury of making mistakes. The practice of guided trial and error no longer has the devotion of time necessary to make it a fruitful process. Testing has imposed timelines in classrooms that have become the most influential goal in planning. There is no longer time to stroll through learning, stopping to smell the roses or make unexpected discoveries. The priority of creating measurable data that prepares students for testing has classrooms racing through curriculums. Taking time away from this forward motion only means you get left behind.

During the week without testing I saw the opposite: I saw messy learning. I saw trial and error. I saw the evidence of as many discoveries as there were dead ends. I also saw collaboration. I watched teams of kids and adults, whole classrooms, and even entire grade levels working together to create projects. There was not pressure to outdo each other or a tension against not moving forward at an appropriate pace. I remember watching the progress of a re-creation of a Native American longhouse going up in one of our main hallways. I walked by students working together with rulers actively debating how long to make a leg for one of the beds that would go inside. Their teacher was around, but they were given freedom to discover how to be successful and learn to complete their task in their own terms. They could, and did, make mistakes. Yet, outside of that week, the collaboration that goes on seems to have a more artificial flavor. Measurable learning is the main ingredient, and it does not allow for messy learning recipes; errors and digressions are not as welcome.

The most damaging effect I see of high-stakes testing is the way antagonism becomes fostered for the students that need the most support from teachers. When I taught first grade two years ago, I was safely distanced from high-stakes tests. In my district in New York they begin in third grade. So, if I had new students enter my classroom mid-year, I did my part to make them feel welcome and to transition them both socially and academically to their new environment. If I observed that they had significant academic needs, I saw how I could help meet them or find support for them, which did happen. Now, as a teacher in fifth grade, a grade in which my students are tested each spring, I try and transition new students the same way I did when I taught first grade. However, to say I approach things in *exactly* the same way would be dishonest of me. The process looks the same; what has changed is the stress that I now have if a student enters my classroom and I observe that he or she needs more academic support than my average population. And, I can tell you exactly where that tension comes from. It is my worry that this student's test scores, which will not have been nurtured by just my practice, will be used as an evaluation of my teaching ability alone. I lament the unfairness of this as I work to transition my new students, but it remains in the back of my mind like a whispered voice, always there when I am quiet.

I truly wish this was not the case, and I offer my own admissions as an example of one piece of an environment that high-stakes testing is creating. I am also a father; I have two school-aged children, and from that role, I recognize that one of the best ways to teach my children something is to guide them as they live the experiences themselves. Lectures only go so far, and nothing beats a real-life example (and you can absolutely forget trying to plan for these, too); there are many moments that turn into teachable moments, but you have to remain flexible and vigilant. In my classroom I feel more at odds with those teachable moments. I use them, but not all of them. There are plenty that slide by because they do not serve to move students forward in the timeline that will have them ready for the tests they need to take in the spring. Therefore, another realm in which I feel antagonism is against my own instincts as an educator. I want my teaching to be messy, and I want to use these teachable moments, which are typically highly engaging, to the fullest, but the ratcheting down of high-stakes, test-based evaluations on teachers is making this a much more difficult task to undertake.

The final relationship that I see bending under the weight of test-based antagonism is the bridge between a school and its community. I am speaking of the parent/teacher relationship. Our week without testing culminated in a visit from the author Matt McElligott. This was an open event to the community, and parents were encouraged to walk through the school to see the fruits of the labors put into the projects created during the week. The enthusiasm was palpable, and the pride in the parents and students alike was very special. Conversely, the highly rigid test-based education system has created a level of efficiency that makes many parents feel left out. Schooling becomes only the job of the school.

From the teacher's perspective this is not as beneficial as it may sound. We rely on support from families to help with homework, create positive academic environments at home, and communicate with us about children. Parents have significant insights that can really help teachers. The view that education is only the job of a school is eroding this collaborative relationship. Further, since teachers are being held more and more accountable by assessments, there is certainly frustration when they perceive parents as not being willing to aid their academic efforts with home support. On the flip side, teachers are blamed exclusively when students fail to achieve. Testing is driving this wedge further down.

My Attempts at Deep Learning and the Resistance of Students to Struggle

I like seeing the big picture. I love uncovering the webs of connection that show the interdependence of our knowledge with our task of survival. I want to know where our systems of language, science, and math come from, and I get excited when I learn what circumstances lead to, or influence, the development of modern ideas. For instance, I teach social studies to all fifth-grade students. We concentrate our year on the development of our country between 1800 and 1900ish. We also learn about our Constitution, Bill of Rights, and the structure of federal, state, and local governments. I love telling students about the Third Amendment, which states that "[n]o Soldier shall, in time of peace be quartered in any house, without consent of the Owner," and that this resulted directly from colonists being forced to house British soldiers during various conflicts prior to America gaining independence (U.S. Citizenship and Immigration Services, 2009). There is a direct connection between an event and law. Too often our history takes on a disconnected feeling. Students seem to think that our government was discovered on a document in a cave somewhere and not that it was the convergence of philosophies, experiences, people, and events. I think when students view our government this way they can see themselves as being part of it. They become more invested. Well, that is my stubborn hope, anyway.

Unfortunately, resistance has sprung up in many of my students to the style of learning that I try to impart. It is not complete, and the level of push-back varies, but I am discouraged nonetheless. And, again, I feel that I can draw a line between this growing resistance and the climate created by high-stakes testing. Producing measurable data requires unidirectional knowledge attainment. *Students must be taught to correctly make their way to a predetermined instructional conclusion in an attempt to produce quality data*. Therefore, conclusions that students make must be limited and approved by whoever is writing curriculums or tests as valid learning. This is like learning with blinders on.

The problem this creates is two-fold: The first deals with instruction. When curriculums are written around this type of single directional learning, they are heavy in fixed answers and not so much in a process or a spectrum of correct options. This is great for testing purposes, but what about the cultivation of thinking? This limits the need to judge an answer correct, or adequate, rather than being told that the answer is right by some sort of authority (usually the teacher). Ironically, I have also found this limitation creeping into teacher thinking as well.

When I started teaching I relied heavily on programs to guide my instruction. For example, for reading, my students read primarily out of anthologies that had teacher additions provided along with them. These resources were packed with prewritten themes and even questions that I could use to stimulate conversations about drawing conclusions, picking out the setting of a story, and many other reading comprehension skills. I needed these as I managed all of the duties that came along with my new teaching assignment.

As I started to gain more confidence as a teacher, I challenged myself to move further away from this safety net. I thought of new ways to instruct my students. (Did you catch that last important bit? I *thought* of new ways to teach.) The materials I relied so heavily upon as a novice teacher became resources for me to draw into my own creative practice. I no longer utilized them like scripts to help me meet educative goals. They certainly helped me a lot, and I still use them from time to time, but I can honestly say that they are not my primary tool anymore.

I work hard to think of novel ways to make learning engaging and relevant to my students as well as to stretch them as learners.

I feel that this approach is threatened by high-stakes tests. I have always kept a close eye on the data that my students produce to see if they are benefitting from my instruction, but I worry that increased reliance on data from testing will necessitate the usage of prewritten curriculums that are aligned with skills and content that are featured on tests. I am worried that my job will no longer allow me to think for myself. I will be forced to teach in one test-guided direction.

This brings me to the second place of friction, my students resisting to think for themselves. No matter how much teachers try to instruct in a thoughtful way, they are ultimately anchored, these days, by high-stakes tests. That means that a significant part of each year is devoted to test prep and taking the tests. Students begin to believe that learning is only a one-way street. Teachers may not teach this overtly, but it is the unspoken theme of all the test-based education work. Students start to become used to taking a single path from a problem or question to a solution. Further, as I mentioned above, they see teachers, not themselves, as the people with the maps. They are less comfortable letting themselves exist in places of dissonance, learning how to judge an adequate solution to a problem or question; too often they figure out that an exacerbated teacher will eventually tell them the directions to travel in, finding the one correct answer, if they convey enough frustration. What I am describing is the amount of resistance that I encounter when learning becomes process driven and not unidirectional. I create learning webs; I pull the veil back from the big picture. This type of learning is more thoughtful, and it is more resisted by students who are so used to learning with blinders on.

One of the most relevant examples I have of this in my classroom involves one piece of my writing instruction. My students spend a lot of time rewriting, or at least that is the design that I have established. They are involved in collaborative groups in which they attempt to hone the clarity and appeal of their writing for an audience (who, by the way, is not just me). This sounds great in theory, but in reality many of my students continue to produce work that is similar to drafts they create the first time. This is especially true of assignments that students produce without any feedback from me. They do what I ask them to do, but they are not as consistent in applying the learning to their independent practice.

I also have my students write comments for each other's pieces. What I have found is that the majority of my students write generic statements about the quality of another student's writing. Their feedback does not even show that they have read the pieces. There are a lot of "This looks great!" or "Add more details." They do not look deeper into the classmates' pieces. They do not latch onto a single interesting sentence and encourage them to fill that out and explore that idea more. I do this when I comment, and it produces better writing, but this also bolsters the notion that I am the final authority on what is *good writing*.

Despite this outcome I keep plugging away. I use student examples and workshop them, with my class, looking for those single bright spots that contain the epicenter of a truly powerful story, poem, or essay. Together we practice writing comments that can really help nurture a piece of writing. I understand the importance of guiding this type of messy instruction; it is a web, not a strait line from problem to solution, which forces students to live on the type of uncomfortable edge that really helps them grow. There are no easy answers, and there are certainly no formulas, but high-stakes tests and the concentration on them are creating more reason for students to not try this type of thinking, even in writing. Why should they? When bland formulaic writing results in good testing scores it makes students feel like they have done enough. The introspective, nonlinear, messy aspect of education goes by the wayside.

Playing in My Classroom

A guitar has hung on the walls of each of my classrooms for several years now. At first I think I brought it in as a way to up my cool factor with my students, but over the last few years my use of the instrument has evolved. On the one hand my guitar serves a practical purpose as I use certain songs to signal different transitions such as lining up or gathering at the front of the room. I can add some soothing background music when students are working quietly, and my guitar accompanies the songs we sing together in class. However, after the No-Testing Week my guitar has taken on a new symbolism for me. It now represents the aspects of education that I worry will be lost amid the current scramble for creating data in all school endeavors.

When I started using my guitar in the classroom I made sure that a direct line could be drawn between its use and some measurable aspect of my instruction. All the songs I brought in were written with the intention of conveying some discrete skill or piece of information, typically related to phonetic development. I also wrote some of my own songs in order to aid students in remembering content in social studies and science. Overall, it was quite successful.

When I switched from teaching first grade to fifth grade in 2010 I knew my guitar use would change. I had a hunch that singing reading songs to the tunes of such childhood classics as "The Wheels on the Bus" and "B-I-N-G-O" would not hold the attention of my new tween audience. As a result, I became more creative with how I used music. I still tried to angle songs toward supporting reading, writing, and other content skills, but I began choosing more updated selections to keep students interested. I think I did fairly well with using various pop and rock songs to teach about topics such as verb tense, point of view in narration, author's voice, and many others; but I still worked hard to explicitly connect my use of music with measurable instruction goals. This lent itself well to certain song and skill pairs, but others were a stretch to say the least.

My guitar continues to hang on the wall of my classroom and I use it daily, although time constraints have changed the role that music plays in my practice. I still try to enrich my instruction with songs that help students remember Social Studies content, but my use of singing and lyric analysis to help instruct has effectively been whittled to almost nothing. In fact, the choice I have had to make this year is between analyzing the lyrics of songs and just singing them. I have gone with the latter. Sure, I still tell my students to follow along with the text even if they are not singing, as a means to assist with reading fluency, but mostly we just sing, and this change in program has not come without side effects of doubt on my part.

The positive effects of using my guitar as part of my teaching are quite evident. For example, each year I am told by students and parents alike that I have inspired someone to dust off and play a guitar that has long sat abandoned in the corner of their house. I also hear yearly of many new guitars that are given to my students for birthdays or holidays. My students are proud to tell me when they are taking lessons. I have many side conversations with kids about where I purchase my guitars, strings, and even where I get my picks. Many guitars have been brought in for me to tune by students who don't know how to do it themselves, and I have listened to my share of original compositions played during show-and-tell times. I know that my instrument has excited and inspired many kids; yet, I have questioned my use of it more and more. Which of these feel-good experiences can I pull measurable data from?

I choose my songs for specific purposes. Granted, most of them have spent time on the radio at some point and they are familiar and fun to sing, but they also are chosen for their messages. If I had to generalize themes I would say that I look for songs that are about valuing oneself, living genuinely and showing compassion to those around you. These are all topics that do not show up regularly on high-stakes tests, I know, but they do allow me to talk with my students about fundamental issues that lie at the heart of success in any endeavor, valuing who they are and valuing their peers and community members equally well.

Yet, I cannot help but think that it is only a matter of time until the data machine chews this practice up and spits it out. How does this part of my instruction neatly meet a standard? What section of future tests will singing help with? In my gut, I believe that using music in the classroom is vital, but will this practice become collateral damage at the hands of the current data-driven movement?

Will the Song Remain the Same?

These days, when I pull my guitar off the wall I occasionally wonder if there will come a day when I no longer have it in my classroom. Sometimes I even wonder if playing to students will take on a new meaning, that of rebelliousness, singing an alternative narrative of how they can and should learn to the powers that be. As long as I teach in an environment where a week without testing is valued and prioritized I do not believe that will be my immediate fate. Still, my heart gets heavy at times while I make music with my students wondering what the future may hold.

Their experience with me helps to soothe my doubts. Whether it is the students who close their eyes and raise their voices, losing themselves in the joy of singing, or the excitement I hear as they belt out the highest notes I throw at them or growl out the most rocking "yeah-eh" that I could wish for, their affirmation bolsters me.

Ultimately, I know I am doing right by them, providing rich content in an engaging way upon which they can draw their *own* conclusions and make their *own* discoveries, free from the confines of measurable data-driven endpoints. Even more affirmation comes in what I see after the singing ends. The kind words and phrases exchanged among my students, or someone reaching out in kindness to connect with one of the loners in our class, is supported and cultivated by the messages we have sung together. We experience a true joy of interacting with the written text and the human experience, artfully learning, uncovering the tradition of living that undergirds all knowledge, and willingly participating without being bribed by a score or a grade.

This is exactly what is being squeezed out of our modern schooling by high-stakes testing. The music produces no data, but it is vital to cultivating the next generation of thoughtful members of our democracy. I believe this wholeheartedly, and I sincerely hope that I will always have the opportunity to take time out of my teaching schedule to sit in a circle with a group of students, and my guitar, and absolutely sing our hearts out. Even better, perhaps my guitar will be allowed to develop as a more central vehicle for my instruction, just like the messy-creative-learning that blossomed out of Poestenkill Elementary School's week without testing. I suppose only time will tell whether or not this guitar-playing teacher's song will indeed remain the same.

Reference

U.S. Citizenship and Immigration Services & United States. (2009). *Civics and citizenship toolkit: A collection of educational resources for immigrants.* Washington, DC: U.S. Dept. of Homeland Security, U.S. Citizenship and Immigration Services, U.S. Govt. Print. Office.

CONCLUSION

Striving toward Authentic Teaching for Social Justice

Lisa William-White

Introduction: Preparing Teachers

> Knowledge need not be regarded as a sacralized text . . . or an inviolate procedure as on the assembly line; nor is it only the more complex, sometimes even creative means to an end as it is in the corporate model. Rather, knowledge and intelligence as free exploration become wings by which we take flight, visit other worlds, returning to this one to *call others* to futures *more life affirmative* than the world we inhabit now [emphasis mine]. (Pinar, 1994, p. 247)

> [Thus,] what schools are *supposed* to do is a complicated question. However . . . there are at least two major purposes to schooling: to educate students in various academic or cognitive skills and knowledge, *and* to educate students in the development of individual and social skills and knowledge necessary to function occupationally and socio-politically in society [emphasis mine]. (Fullan & Stiegelbauer, 1991, p. 14)

What does it mean to *prepare* emergent K–12 public school teachers in the United States in the current historic moment? More specifically, what does *preparation* mean in an age where we face economic uncertainty in the United States and abroad; where American students hear about, observe, and/or *experience* the effects of war (Iraq and Afghanistan)? More than two million troops have been deployed to Iraq and Afghanistan; 55% of the force is married and 40% have two children; and 63% of families live in more than 4,000 communities nationwide (Strengthening Our Military Families, 2011). What should preparation look like as we observe local and worldwide protests of the "99%" to advocate for economic equality and fairness, when the median income level has declined for

U.S. families; and roughly 20.5 million people have income below the poverty threshold (Watt, Proctor, & Smith, 2011)?

What does it mean to *prepare* emergent teachers in a country where we comprise 5% of the *world* population, yet 25% of the world's prison population? What does preparation mean when we spend "$70 billion annually" to incarcerate adults; confine youth in detention centers; and pay for personnel to supervise 7.3 million individuals on probation and parole (Hawkins, 2010; Misplaced Priorities, 2011)?

What does it mean to *prepare* emergent teachers in an era where we bear witness to anti-immigrant discourses and policies; where we see (*or even know*) scores of people who live in poverty (Measuring Child Poverty, 2012); or where there is widespread bullying of children and youth in schools and communities (From Teasing to Torment, 2005)? What does *preparation* mean in a country where we have championed education reform since the 1950s; where we extol the importance of literacy and *critical thinking*; and yet, we further *prescribe* what constitutes appropriate *knowledge*, including what *content* teachers must teach (Common Core State Standards Initiative, n.d.)? And, what does this all mean in an era of education *deform* (Pinar, 2012)—a time of shrinking state budgets, eroding of educational enrichment opportunities for children and youth, rising tuition costs in universities, and where democratic learning spaces in higher education are further undermined by business models for educational decision making?

From my standpoint as a *teacher* and a *teacher educator*, the training of emergent teachers is weighty, a vocation that must be shaped by critical consciousness about these aforementioned issues combined with a moral imperative to serve the poor and culturally and linguistically diverse communities (William-White, Muccular, Muccular, & Brown, 2013). Moreover, through the study of real-world issues lay opportunity to examine the intersections of geography, history, mathematics, economics, and civics, to name a few of the opportunities for academic disciplinarity. Thus, competence means more than meeting federal or state definitions of *teacher-readiness* to teach; it certainly is more than the acquisition of knowledge, skills, and dispositions created by national organizations and accrediting bodies that govern teacher education.[1] Rather, competence entails:

> [D]eveloping . . . personal and interpersonal awareness and sensitivities, learning specific bodies of cultural knowledge, and mastering a set of skills that, taken together, underlie effective cross-cultural teaching. (Moule, 2012, p. 5)

Culturally competent emergent educators embrace interdisciplinarity and utilize students' knowledges and interests to examine subject matter knowledge and its relationship to the broader sociopolitical milieu, as opposed to embracing reductionist teaching emblematic of top-down schooling reforms and the standards regime. Lipman (2009) shares:

> Under the rubric of standards, the policies impose standardization and enforce a language and cultural assimilation to mold the children of the increasingly linguistically and culturally diverse workforce into a more malleable and governable source of future labor. This is a system that treats people as a means to an end. (p. 371)

Competent teachers disrupt the "pedagogy of poverty" that is replete in many poor schools (Kohn, 2011) where intellectual curiosity remains untapped due to the adoption of scripted curriculum to meet federal standards. Instead, these educators engage in an ongoing commitment to create "caring learning communities," while working unrestrained for social justice within the culture of the school (Teel & Obidah, 2008, p. 3). These educators are also aware of, and understand, the imperative of embodied praxis (see Figure C.1).

Figure C.1. Diagram of a Process-Based Approach to Curriculum Inquiry From William-White, L., Muccular, D., Muccular, G. & Brown, A. (2013).

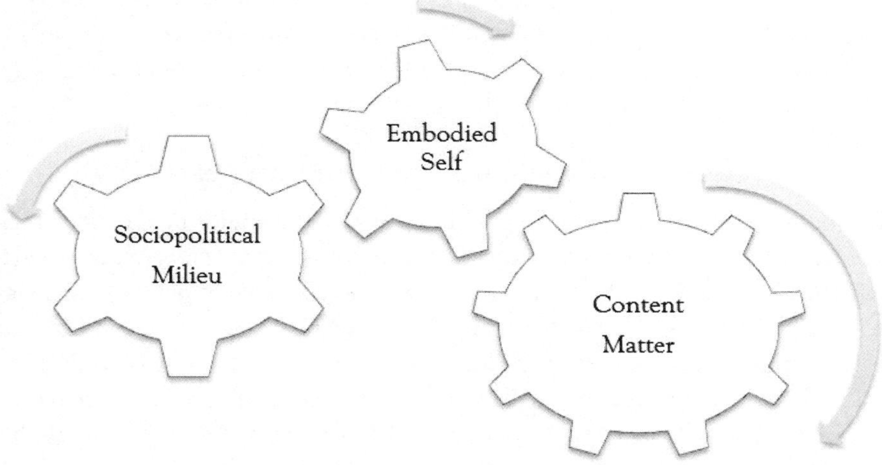

Additionally, these educators possess knowledge of and the ability to understand and explain the evolution of U.S. public education, and the influence of major educational legislation—*Brown vs. Board of Education*[2] and *Lau vs. Nichols.*[3] They possess knowledge of the agency and activism that arose from the minority groups who used their positionality to push for educational equity and understand the link between schooling and greater societal mobility. They also understand the trajectory of the federal role in public education and how the Elementary and Secondary Education Act led to No Child Left Behind (NCLB).[4] Furthermore, they comprehend the purpose of programs such as Title 1 (assistance to the disadvantaged) and Title VII (Bilingual Education Act); and they draw upon these foundations to understand the politicized teaching landscape in which they are positioned. Ideally, these educators also draw upon critical pedagogy (Freire, 1971), critical race theory (Ladson-Billings, 1998), and sociocultural learning theory as relevant frameworks (Lave, 1991; Moll et al., 1992) to strive toward transformative teaching. Moreover, they challenge and subvert schooling practices focused merely on testing and grading—mechanisms that promote a highly racialized discourse of deficits with regards to children (Lipman, 2009) while negating issues that mediate the lives of people within oppressed communities.

Educators of this ilk engage their students in a "discourse of democratic public life" and critique (Giroux, 1997, p. 241), which inverts the corrosive focus on schooling as merely an apparatus for the free market economy and social reproduction. These critical educators view schools as public spaces to examine hegemonic power relations in American society, rather than embracing the right-wing discourse of competition and school choice. Furthermore, they are cognizant of the fact that this same discourse, birthed during the Reagan administration, is relied on today to extol the *virtues* of individual hard work and merit, minimizing the material effects of the unequal distribution of resources in people's daily lives (Giroux, 1997) and within schools (Kohn, 2011).

Teacher Competency in the National Realm

The notion of teacher competency as codified by law and educational policy initiatives juxtaposes my aforementioned view of emergent teacher competency. Under President George W. Bush, NCLB required states to assess students' proficiency on basic skills to work toward closing the achievement gap, with the ultimate goal of ensuring that American children are competitive with other industrialized nations.[5] At the same time, all teachers must meet the federal designation of "highly qualified" to teach in public school classrooms (Meeting the Highly Qualified Teacher Challenge, 2003). Thus, under the terms of NCLB, highly qualified teachers (HQTs) needed to meet basic requirements for teaching competency (see Table C.1).

States were required to outline how they planned to meet a variety of teacher quality goals set by the U.S. Department of Education, one being that 100% of all public school classes be taught by HQTs[6] (Highly Qualified Teachers, 2008). This mandate placed greater pressures on states and local school districts to ensure compliance with the law. And high teacher turnover in schools is particularly dif-

Table C.1. Federal Requirements under NCLB

- Hold at least a bachelor's degree from a four-year institution
- Hold full state certification
- Demonstrate competence in their subject area
- Newly hired elementary school teachers working in core academic areas must:
  a. pass a rigorous state test of subject knowledge and teaching skills in reading/language arts, writing, math, and other areas of the basic elementary curriculum.
- Newly hired middle school and high school teachers in core academic areas can demonstrate their subject-matter competence by:
  a. passing a rigorous exam of their content knowledge;
  b. majoring in their subject as an undergraduate;
  c. earning a graduate degree in their subject; accumulating the coursework equivalent to an undergraduate major;
  d. or attaining an advanced certificate or credential.

ficult for the culture of schools that traditionally serve many low-income students and students of color (Lipman, 2009; Ronfeldt, Loeb, & Wyckoff, 2012).

The notion of teacher accountability and student achievement continues to be the focus of discourses in the Obama administration—merit pay for teachers who raise student achievement on standardized tests; the bedfellow to this idea is more systematic and intensive approach to the assessment of emergent teachers through their preparation and training programs. For example, a report commissioned by the Center for Education Progress posits that the current educational system in the United States does not hold teacher education programs accountable for candidate learning and meeting high standards in K–12 public schools (Crowe, 2010). These discourses have gained traction in educational reform discussions, particularly as greater dialogue ensues related to alternative teacher preparation pathways such as Teach For America and the New Teacher Project, and notions of instructional innovation in meeting the demands of K–12 students.

In fact, there are those who claim that university-based teacher education programs are "unwilling" to address issues of teacher competency through failed efforts at *quality control* (selective recruitment into programs; carefully constructed and monitored clinical experiences for teacher candidates; and program evaluation focused on important outcomes) to ensure quality graduates are able to raise achievement (Crowe, 2010). Yet, research on teacher preparation and professional development is a "long way from the stage of converging evidence and professional consensus" about the quality of teachers' training in varied programs (Meeting the Highly Qualified Teacher Challenge, 2003, p. 40), and there is little research on the interventions that raise effectiveness for all teachers. The subtext about interventions is wrought with problems as well, as the notion of *effectiveness* is a construct worthy of further examination (for instance, effective to whom and based on what standard?). Nonetheless, Crowe (2010) suggests that systems be established that meet the following goals for teacher competency and, by implication, university-based teacher education programs:

- Tie K–12 pupil learning outcomes to preparation program graduates and hold the programs accountable for teacher effectiveness
- Begin to implement high-quality observational assessments of classroom teaching by supporting efforts to link these assessments to student achievement and by developing rigorous training for classroom observers to ensure reliable assessment findings
- Employ current state data systems to track the teaching persistence rates for graduates of every program, and use the findings as a public disclosure measure
- Implement feedback surveys of preparation program graduates and their employers using state education, labor department (or state insurance department), university and school district data systems

In all, social justice teaching is not a consideration in the dominant policy discourses on emergent teacher quality or competency. Moreover, critical questions about the nature of teaching are not raised particularly with regards to what actually happens *in* classrooms (Kohn, 2000), or with regards to the varied ways that learning is actualized in classrooms; or, even what role students should play in their own learning. For instance, those of us who teach know that there is no one-size-fits-all intervention or approach to meeting the diverse and divergent needs of students. We know that drill-and-kill activity and prepackaged lessons are the antithesis to motivating learners who need differentiation to demonstrate growth and understanding. Instruction absent of this pedagogic knowledge is irresponsible at best and is certainly detrimental for students. Consequently, there needs to be an elevated, alternative dialogue about notions of teacher competency and effectiveness that demand that teachers inform and drive policy discussions as opposed to merely implementing bureaucrats' visions of what *should* take place in classrooms.

Yet, conservative education critics view such pedagogic considerations and differentiation as antithetical to school reform and national educational progress. This means that business thinking and efficiency models shape discourses on curriculum and instruction—which approaches will yield "success" for schools (Lipman, 2009) and allow the United States to maintain its competitiveness in the world economy. And much of this discourse, such as the move toward instituting a common curriculum, harkens to a national agenda of cultural indoctrination and assimilation into a national identity, but it also serves to undermine democracy in a multicultural and multilingual America. Giroux (1997) well articulates the motives of conservative critics of multiculturalism:

> "Whose experiences, histories, knowledge, and arts are represented in our educational and cultural institutions? How fully, on whose terms, and with what degree of ongoing, institutionalized participation and power?" . . . in a democratic society should subordinate groups attempting to fashion a pedagogy and politics of inclusion and cultural democracy be derisively labeled as particularistic because they have raised serious questions regarding how the canon and public school curriculum work to secure specific forms of cultural authority or how the dynamics of cultural power works to silence and marginalize specific groups of students. (p. 244)

In sum, public school and higher education curricula are sites of ideological struggle, one where the modes of cultural transmission complicate the curricular conversation and reduces the possibilities of social justice teaching.

Impact of Accountability within a Teacher Education Program

Cultivating radical and progressive educators who are committed to social justice teaching has become increasingly challenging within this age of accountability, particularly when training candidates within a context with competing objectives, outcomes, and assessment goals. For example, charged with meeting the demands of NCLB and state reforms focused on accountability, the community partners (school administrators and cooperating teachers) where my department historically has placed candidates for their field experience in student teaching face increased scrutiny related to raising achievement based on standardized tests (Berta-Avila & William-White, 2010). Consequently, pressures to raise student achievement and meet adequate yearly progress based on the state's system for school accountability would temper many administrators' willingness to accept student teachers in their sites. In fact, some administrators have been outright opposed to having student teachers due to the need to be in compliance with state law (and parents) to ensure that HQTs are responsible for daily instruction.[7]

In addition, our partner schools (typically Title 1 schools) heavily focus on benchmark assessments and standardized testing, which includes increased utilization of scripted curriculum and pacing guides (to raise student achievement to adequate "proficiency"— levels as defined by the state) (McNeil, 2009). Thus, candidates placed in these sites are given little latitude to deviate from skills-based lessons that are often prepackaged. This context proves problematic for many reasons, most significantly, my department's social justice mission and teacher performance expectations (TPEs; to be discussed later).

Added to these challenges is also the criterion utilized for determining emergent teacher competency. In 1998, the state of California passed legislation (SB2042) that would require teacher candidates enrolled in credential programs to successfully complete a teaching performance assessment (TPA) to obtain a preliminary teaching credential. Programs had two options: they could administer the TPA designed by the state in consultation with the Education Testing Service (ETS), or they could develop their own TPAs, provided they met the state's Assessment Quality Standards.[8] During this time, an alternative performance assessment designed and piloted in the spring of 2003 by a consortium of preservice teacher preparation programs throughout the state[9] was piloted by varied programs in my college in 2005. The assessment, called the Performance Assessment for California Teachers (PACT), is a high-stakes standards-based performance assessment designed to measure preservice teacher learning. Berlak states, "the assessments are required by law to be *performance-based,* that is, they are expected to assess evidence from teachers' actual teaching practice" in real world contexts (Berlak, 2008).

The PACT, adopted college-wide, draws on varied data sources to measure candidates' ability to plan, implement, assess, and reflect on an integrated series of lessons that are intended to facilitate quality instruction for all learners. This assessment of teacher competency is broken down by content area, framed within knowledge and practices that are valued within the structure of the discipline. Berlak (2008) explains:

> PACT assesses two types of "performance," the *Teaching Event* (TE) and several *embedded signature assignments* (ESAs). The Teaching Event is a fifteen-minute video-taped lesson taught by the candidate during the final semester of the certification program, accompanied by a portfolio of approximately fifty pages which includes multiple sources of data related to the TE: teaching plans, teaching artifacts, student work samples and written reflections and commentaries. The guidelines specifying the elements of the portfolio are laid out in great detail in a 47 page Student Handbook. . . . The signature assignments are course assignments in specified content areas graded by the teacher of the course using state-approved criteria.

English language arts, for example, requires that candidates select a learning segment of three to five hours of instruction that supports students in developing an understanding and interpretation of complex text and in creating a written product responding to text. Likewise, history/social science candidates are required to select a learning segment of three to five hours of instruction that help students use facts, concepts, and interpretations to make and explain judgments about a significant historical event or social science phenomenon. The assessment plans for the learning segment within both of these disciplines are focused on the creation of student products that reflect the extent to which the class met predetermined standards/objectives that can be collected as evidence and measured. This process would then include an analysis of student achievement based on what students "did or did not understand" (Performance Assessment for California Teachers, 2012), emblematic of our "romance" with science vis-à-vis quantifiable evidence (Berlak, 2008). Not surprisingly, the core assessment for all disciplines is the application of subject-specific pedagogical knowledge that is often skill-based with finite assessment goals. Failure to pass the PACT means that candidates are not able to obtain a credential, even if they have passed all of their courses and have passed their student teaching field requirements.

In a study conducted in 2008–2009, candidates in our Single Subject credential program had passed their PACT, yet they still lacked the ability to develop culturally responsive units/lesson plans that reflected their students' needs, interests, and lived experiences. Berta-Avila and William-White (2010) explain:

> [W]hat gives us pause is the knowledge the dominant measure of teacher competency through the PACT does little to examine the lens of what is being taught and for what purpose. Additionally, the privileging of a skills-based approach to teaching such as lesson planning, instruction, and assessment within most teacher education programs, along with the PACT assessment, encourages a behavioral approach to teaching, the antithesis of praxis that yields little examination of the politics of teaching, and the selection of strategies, method and content. (p. 418)

Consequently by the 2009–2010 cohort, an increasingly prescriptive (albeit counterintuitive) approach to teaching our candidates occurred, whereby we required them to choose unit themes based on the real circumstances of students' lives (Berta-Avila & William-White, 2010). Yet, our candidates struggled with this process due to the tendency to privilege the subject matter content and standards over the needs of their students. Although candidates were able to describe their own schooling experiences and draw connections to examples of social inequity, their attachment to their subject-matter content hindered them from viewing their students as the center and purpose of their curriculum.

During the 2010–2011 academic year, I continued to collect data (lesson/unit plans and reflections on teaching) from cohort groups and found that (1) attempts to intensively examine candidates' curriculum development and instructional planning were frequently usurped by candidates' focus on the PACT assessment; (2) attempts to engage candidates in dialogue and reflection about their own learning, which included offering my evaluation of their efficacy related to department-level teacher performance expectations, was met with resistance; and

(3) though candidates expressed belief in, and desire to engage their learners in, transformative teaching, these efforts were subsumed by increasing pressure to focus on increased accountability expectations from public school placement sites *and* pressures to pass the PACT. In fact, candidates vehemently expressed anger about the structure of the credential program including the fact that faculty still expected them to read and submit coursework while they were attempting to work on their PACT and meet their student teaching obligations. For a few candidates, the unwillingness of faculty to accommodate their course load[10] and stress levels was the antithesis of being student-centered. This was a major tension, as candidates' adherence to due dates was problematic. Not only that, submission of required field assignments for in-class discussion, analysis, or utilization to build toward other learning tasks left much to be desired.

In sum, after much frustration and reflection on the varied ways that testing and grading usurped a focus on social justice teaching efforts, major curriculum revision was undertaken for my multicultural education course. My attempts to counter these aforementioned challenges, to facilitate more engaged praxis, and to elicit more critical self-reflection from candidates about the politics of teaching and teacher praxis was the major focus of my curriculum-as-planned for the incoming cohort for the 2011–2012 school year.[11] Revision of the curriculum included a more intensive focus on dialogue and authentic assessment that provided a counter-script to the PACT and the national focus of teaching competency under NCLB. Most importantly, I desired for my candidates to have more ownership *of* their learning by exalting more process-based learning attempts in their field practicum while minimizing a focus on producing artifacts for the purpose of grades.

Critical Pedagogy and Authentic Teaching for Social Justice: Description of Scaffolds for Emergent Teacher Competency

Several assessments were developed and utilized in my course, described as follows:

> A critical analysis of the purposes and processes of public schooling. Examination of the sociopolitical relationship between public schools and society, and between educational theory, culture, community and educational practice. Social, cultural, historical and philosophical foundations of education; learning theories; and ethno-cultural, social, emotional and cognitive development. Candidates examine their attitudes regarding gender, sexuality, race, social class, language, and ability. Candidates develop a philosophy of education for our multicultural and democratic society. Lectures, discussions, small group work, simulations, and field tasks comprise this course.

In the fall semester, my course begins by exploring and analyzing candidates' personal knowledge and experiences by asking them to reflect on their learning trajectory and views of the purpose of schooling—their currere (Pinar, 2012). Throughout several sessions intended as community building opportunities, candidates journal, dialogue, and have ample opportunities to share their perspectives about U.S. education in general and about their specific personal experiences within K–12 public education. A concept map is used as a graphic organizer to aid candidates' thinking.

This preliminary task is followed by a community-building process called "Long Introductions" where I introduce myself and speak about my background; major events that shaped my consciousness (related to race, class, gender, ability, sexual orientation, religion); and other background information that I deem as pertinent in my schooling and professional journey.

I model this process of storytelling, speaking anywhere from 15 to 20 minutes so that students understand the importance of subjectivity (values, beliefs, worldview, and educational philosophy, to name a few) in the learning context, including how subjectivity shapes praxis. Each candidate then is given an opportunity to share his or her journey.[12] This activity takes place on the first day of class and extends into the next course session if needed—as long as it takes to give all participants an opportunity to share. Active listening and learning is the goal, followed by a few minutes for class members to ask clarifying questions or to make connections to each other's stories. Typically, this process proves to be quite emotional for many candidates, as it often is the first time that they have been guided through a learning activity where they are encouraged to think about and share their intimate, and not uncommonly painful, experiences with schools. A candidate who teaches history shared the following:

> I did not see non-white's stories and contributions represented in the curriculum. I felt that what I was learning was untrue. I would always ask my mom to tell me more about my people and other people's history and how they contributed to the United States. Of course, she only knew about my people's history, but after understanding the Hmong people's contributions to the United States, I understood why my parents came to the United States. With the knowledge my mom shared with me, when I took World History in tenth grade, I was excited to learn about the Vietnam War because I wanted to see and hear of how the Hmong people and General Vang Pao assisted the CIA. I wanted my teachers, who I considered the experts, to validate my mom's knowledge, to validate my people's contribution. There was nothing. Nothing happened. Since that day, the concept that white people only want to brainwash non-white people so we can be their fuel for superiority grew in me. . . . My experience as a second year undergraduate student only strengthened that concept. I took a history course in Asian Civilization and history repeated itself. . . . The professor did not

> mention one thing about Hmong people's contributions; we were only referred to as "other minority Asian groups" who helped block the Ho Chi Mien Trail.

This activity is one of several where I elevate the curriculum-as-lived and work alongside candidates to connect themes from their lives to the course goals (curriculum-as-planned), particularly the sociopolitical foundations of U.S. education—*education historically for whom and why?* For those who experienced marginalization or invisibility within the curriculum, this experience had students think about the textbooks they use in their placements and discern in what ways their students' cultures are represented in those texts. We begin to ask questions and chart out what we know about the education history of varied ethnic, cultural, and linguistic groups in the United States (drawing from the diverse cultural groups represented within our cohort). Another candidate shared the following:

> Teachers should always care about all their students without excluding any of them. I, for example, didn't feel valued by my English teacher when I was in high school. From my point of view, she didn't care what the Mexicans did in her class. She did not seem to value any other culture but hers. This is exactly what teachers should not do. Another thing that got my attention was the fact that some staff members and the principal looked at African Americans as very low skilled, unmotivated, and as trouble makers. . . . We can still see this happening in different schools. . . . This is called racism.

This activity provides another opportunity for candidates to share their personal and historic knowledge, allowing for further inquiry into the structural and historic conditions that construct and reproduce inequities for varied groups within U.S. society. For example, we examine the intersection between economics and race, such as how the federal government's economic interests produced Native American boarding schools, and how economic interests created the need for cheap labor, which in turn institutionalized the system of slavery that deeply impacted African American lives, particularly issues of exclusion.

Framing issues in this manner enables our class to explore and examine the foundations of U.S. education from both a systems perspective and from a subjective standpoint. We draw upon scholarship, narrative texts, film, and alternative media to humanize the curriculum, and at all times candidates are encouraged and expected to speak, journal, and react to what they are exploring as we move through our learning trajectory as a classroom community. Furthermore, theoretical positions that we explore such as Freire's (1971) critical pedagogy texts and Banks's (2008) and Sleeter and Grant's (2003) multicultural education frameworks allow us to further examine how varied philosophies of schooling have contributed to the teaching landscape. Samplings of seminal texts for the course are listed in Table C.2.

Table C.2. Selected Sample of Course Materials

Sample of Core Readings and Films
• Freire, P. (1998). *Teachers as Cultural Workers.* Boulder: Westview Press. • Freire, P. (1971). *Pedagogy of the Oppressed.* New York: Seabury. • Banks, J. A. (2008). *Introduction to Multicultural Education* (4th ed.). Boston: Allyn & Bacon. • Robert, J. (2005). *The Heart of Whiteness: Confronting Race, Racism and White Privilege.* San Francisco: City Lights. • Schultz, B. D. (2008). *Spectacular Things Happen Along the Way.* New York: Teachers College Press. • Teel, K. M., & Obidah, J. (2008). *Building Racial and Cultural Competence in the Classroom,* New York: Teachers College Press. • Sleeter, C. E., & Grant, C. A. (2003). *Making Choices for Multicultural Education: Five Approaches to Race, Class, and Gender.* New York: John Wiley. • Moll, L. C., Amanti, C., Neff, D., & Gonzalez, N. (1992). Funds of knowledge for teaching: Using a qualitative approach to connect homes and classrooms. *Theory into Practice, 31*(2), 132–141. • Murrell, P. C. (2002). *African-Centered Pedagogy: Developing Schools of Achievement for African American Children.* New York: SUNY Press. • Lave, J. (1991). Situated learning in communities of practice. In L.B. Resnick, J. M. Levine, & S. D. Teasley (Eds.), *Perspectives on Socially Shared Cognition* (pp. 63–82). Washington, DC: American Psychological Association. • Ladson-Billings, G. (1995). Toward a theory of culturally relevant pedagogy. *American Educational Research Journal, 32*(3), 465–491. • Ladson-Billings, G. (1998). Just what is critical race theory and what is it doing in a "nice" field like education. *International Journal of Qualitative Studies in Education, 11,* 7–24. • Ladson-Billings, G. (2000). Fighting for our lives: Preparing teachers to teach African American students. *Journal of Teacher Education, 51*(3), 206–214. • Documentary film series—*Race: Power of an Illusion* produced by the California Newsreel.

This initial framing provides a springboard for the introduction of the major course assessment tool that is used as an end-of-semester candidate self-assessment (see Table C.3).

Candidates were required to keep a portfolio of their planning in their field placement (lesson/unit plans, assessment tools, supplemental resources, student work samples, parent communications, etc.). Then, using varied artifacts per TPE, candidates describe in a written reflection paper[13] how their praxis reflects the department's TPEs.[14] The portfolio assessment relies on the candidates' narrative and artifacts to gauge authentic purpose for planning and teaching, while enabling me to understand the candidates' instructional choices and rationale for instructional content. More so, allowing candidates to construct and prepare their own portfolios using the guide provides a common language for us to dialogue about the instructional efforts used in field placements. The portfolio evidence

Table C.3. Department Teacher Performance Expectations (with examples to further thinking about praxis)

Department TPEs Candidates will use the Portfolio Assessment as a holistic guide to self-examine performance in each section.
Advocate a social justice perspective within classroom, school, community and/or political contexts (as evidenced by philosophy statements; lesson/unit foci; communiqués with families, colleagues, students, and communities; other documentation of involvement with forums or events). • Examines effective ways to include cultural traditions and community knowledge, values, and resources in the instructional program of a classroom as reflected in lesson/unit plans • Draws upon students' needs as the focus of instructional decision making; there exists a direct connection between Classroom Portraits and other data and lesson/unit plans • Utilizes Multicultural Education Model(s) and/or Critical Pedagogy as the framework for lesson content, purpose, processes, and praxis • Identifies cultural and/or community "funds of knowledge" and integrates knowledge, interests, or expertise into curriculum • Understands theory and research specific to learning (cognitive, affective, and psychomotor), and its relationship to learning and motivation • Examine one's classroom through the lens of language backgrounds, race, class, gender, sexual orientation, ability grouping, with a focus on equity within social and academic foci • Introspective and self-reflective about one's behavior through the lens of language backgrounds, race, class, gender, sexual orientation, ability grouping, with a focus on equity within social and academic foci • Creates and chooses instructional material (auditory and written products) that is free of discriminatory language and bias • Recognizes the relevancy (depth and breadth) of historic and legal policies on education for diverse groups and its impact on curriculum mandates and pedagogical decision making
Use and further develop students' cultural funds of knowledge, bilingualism, biliteracy, and bidialectalism (as evidenced in your processes, methods, scaffolds, rubrics, and student artifacts). • Recognizes students' specific language needs or learners (as evidenced from Classroom Portraits, assessments, inventories, etc.) and utilizes this information to inform classroom planning and learning activities • Utilize knowledge of students' strengths and talents to support appropriate contexts for learning

- Assists students with accessing needed resources for learning
- Candidate's instructional decision derives from knowledge gained in practice with reflection
- Utilizes the history, legal, policy of second language acquisition/CELDT; bi-dialectalism; cultural advantages of bilingualism; advocacy; metalinguistics to support instructional decision making (evidenced in lessons' purpose, language objectives and/or multicultural objectives, and classroom activities
- Utilizes authentic assessment; culturally diverse texts; academic literature; SIOP and ELD methods; adaptation of adopted curriculum; utilization of contrastive language analysis; and varied developmentally appropriate methods and strategies (TPR; basic skills, interactive journals, wordless/ABC books, SOLOM) to make content comprehensible to students
- Produces lessons that are supportive of language development, including L1
- Demonstrates knowledge of K–12 content and ELD Standards

Lead students to achieve at academically high standards across the core curriculum.

- Effectively teaches diverse students by increasing one's knowledge and understanding of the background experiences, languages, skills and abilities of student populations; and by applying appropriate pedagogical practices that provide access to the core curriculum and lead to high achievement for all students. Evidence such as assessments, rubric criterion, artifacts, scaffolds, and reflection on varied assignments utilized within the teaching context.
- Expresses significant insights into the learning process with strong emphasis on analyzing data about student learning, as evidenced through reflection on classroom portrait data and other assessments utilized within teaching context.
- Self-reflects by analyzing what went well with lessons and what did not, in reference to lessons and rubrics (supplied in portfolio)
- Provide students with opportunities to engage in extracurricular activities
- Motivates students to acquire important core knowledge
- Candidate is able to tap into students' background knowledge
- Increases his/her knowledge and involvement in civic responsibilities and opportunities.
- Knowledge of pedagogy reflects research on effective instruction for culturally and linguistically diverse students
- Lesson(s) reflect teaching processes that provide a positive learning environment (respectful, interaction, safe)
- Upholds professionalism in their teaching and collaboration efforts within the school site, as evidenced by documented feedback from CTs, administrators, parents, and/or students
- Candidate's instructional content has powerful effect on student performance as evidenced through assessment of student work samples

- Candidate recognizes and utilizes the social and cultural capital in students, families, school personnel, by integrating these strengths in curricular decision-making
- Candidate maintains appropriate developmental expectations for student writing, academic language usage and academic work
- Candidate provides a safe/civil environment to discuss divisive issues (issues which challenge one's personal beliefs)
- Candidate's field placement behaviors support democracy (field trips, tutorial programs, home visits, observations of school board); political and higher levels of participation in the field of education
- Candidate implements culturally relevant pedagogy for high achievement
- Candidate creates instructional scaffolds for differentiation
- Candidate taps/builds upon prior relevant knowledge connected to instruction
- Candidate identifies cultural and community funds of knowledge and integrates into curriculum

---

Guide students to explore issues of prejudice towards people of different races; ethnicities; socioeconomic classes; language and language varieties; abilities and disabilities; gender and sexual orientation.

- Candidate recognizes and minimizes bias in the classrooms and creates an equitable classroom community that contributes to the physical, social, emotional, and intellectual safety of all students, as evidenced through personal philosophies, classroom syllabi and rules, correspondence to parents and families, etc.
- Candidate systematically examines his or her stated and implied beliefs, attitudes, and expectations about diverse students, families, schools, and communities, and applies pedagogical practices that foster high expectations for academic performance from all participants in all contexts.
- Candidate is able to frame issues based on democratic principles in planning, reflection, and assessment.
- Candidate can write a coherent lesson plan reflective of Multicultural Education tenets.
- Candidate demonstrates knowledge of Critical Pedagogy, Multicultural Education tenets and can apply tenets in instructional decision making.
- Candidate describes race, class, gender, sexual orientation, ability grouping, achievement gap, democracy, equity vs. equality.
- Plans social and educational activities beyond merely a "celebratory framework" for promoting multiculturalism (such as for Black History Month, Cesar Chavez Day, Women's History Month, etc.).
- Candidate knows how to differentiate and adjust instruction to meet the diversity of students and learning objectives.

- Candidate plans activities that indicate an understanding of the multiple learning styles and/or processes with strong emphasis on analyzing data about student learning, as evidenced from developmental, linguistic, and sociocultural data collection.

Promote school transformation toward equity and social justice on multiple levels.

- Candidate provides evidence of introspection and open-mindedness with stress on self-improvement and instructional process.
- Candidate analyzes data about the needs or interests of diverse populations about student learning.
- Candidate's artifacts include samples of teaching that engage students in transformative learning.
- Candidate is a thoughtful and reflective practitioner that draws both on classroom data and theory.
- Candidate displays appropriate professional behavior, reflective of program values.
- Candidate works cooperatively with diverse groups.
- Candidate has gained deep-level insight into teaching and learning through field experiences and coursework.

must demonstrate how candidates engage learners in the interdisciplinary study of social justice issues (to be discussed) that are seen as vital to our department's goals for teacher certification and competency beyond the college's focus on accountability. Furthermore, this process of assessment reclaims our classroom as a site for talk about "humanity, difference, democracy, culture, thinking, personal meaning, ethical deliberation, intellectual rigor, social responsibility, and joy in education"—offering a challenge to the dominant discourse of teacher education as merely job preparation (Lipman, 2009, p. 371), and where competency is equated with providing a service in the market economy of education.

Additional signature assignments briefly discussed below are intended as scaffolds that are designed to help candidates meet the teacher performance expectations outlined in the portfolio self-assessment.

I-Search Paper

The I-Search is a unique form of the traditional research paper, written in the first person. The first-person perspective reinforces and privileges the importance of one's personal standpoint (usually reflective of either the critical or interpretive research traditions), and it decenters third-person point of view or what researchers might call positivist forms of knowing. The paper is intended as a process-based assignment that documents candidates' search for increased knowledge and information about a given topic. Candidates are charged to examine the political, pedagogical, and philosophical discourses and tensions that shape K–12 schools, teaching, and learning, particularly for the poor and culturally and linguistically diverse populations (CLD). Hence, as candidates read varied texts, watch films, and engage in discussions in class, they are continually asked: "What are the implications (political, pedagogical, and philosophical) for teaching culturally and linguistically diverse students?" The requirements are detailed in Table C.4.

Table C.4. I-Search Paper Expectations

Description of research paper sections, which are developed over the course of six weeks.
1. What I imagine, assume, or know about the political, pedagogical, and philosophical discourses and tensions that shape K–12 schools, teaching, and learning: • Section must be at least 2 pages and should comprise knowledge possessed prior to any research and data collection. This section can express your philosophy of teaching and current perceptions and value system about teaching and learning. 2. What I want to know and why: • Section must be at least 2 pages approximately and should reflect your preliminary questions, your curiosities, and/or your preliminary thoughts about the learners with whom you are placed this semester and the political, pedagogical, and philosophical discourses and tensions that shape their learning context. Brainstorming questions would enable you to inquire about these discourses and tensions. These questions are intended to help you think about issues that can/will impact your teaching and students' learning. 3. What I discovered: • Section must be at least 5–7 pages, utilizing APA style (See apastyle.org for models of parenthetical citations and for your bibliography). You can **either** use footnotes or endnotes for source information **or** create a Reference page for this section. • It would be wise and highly suggested that you begin to seek out information from professional organizations in your field.[15] This will enable you to share ideas and insights about the teaching and learning context from a content-level perspective. • Consider what ideas and insights you are gaining from readings in your methods courses, as well as from readings from within and beyond your courses. • Share insights gleaned from data obtained about your learners from your Classroom Portrait data collection. • Interview professors, field supervisors, cooperating teachers, parents, and even students to ask them about their perceptions of the teaching and learning context. Take notes, collect data, and/or audiotape (if need be) to assist you in preparing your paper. You may even consider developing a

survey or questionnaire to solicit information from students or parents.

- In this section, you should compare what your preliminary thoughts were to what was discovered from your newfound insights (related to the political, pedagogical, and philosophical discourses and tensions that shape K–12 schools, teaching, and learning).
- The paper must include all relevant field notes from your Search in an Appendix section.

4. What I have gained:
    - Section must be at least 2–3 pages.
    - Conclude your paper with discussion of how this information will inform your action plan for your lessons/unit design.
    - Does this information provide any big ideas about possible unit plan or lesson ideas to broach in your classroom?
    - How will this information shape your curriculum decision-making and instructional planning? Situate your thinking and ideas within the discourses about critical pedagogy, critical multicultural education, and social justice education.

Classroom Portrait

The Classroom Portrait is another tool designed to aid candidates' development of cultural competence, particularly how subjectivity provides windows and insight for informed instructional planning. To facilitate this process, candidates' inquiry is organized around three sections: the schooling context, ethnographic and narrative research methods for data collection; and a synthesis and reflection section. First, students are asked to describe general demographic information such as the grade taught, the degree of tracking or ability grouping in the course, and varied expectations that might impact candidate's planning or delivery of instruction.

Second, students examine and practice varied research methods (e.g., informal or formal observation, one-on-one and focus group interviews, surveys, and journals) that will support ways to look at data to inform their understandings of the schooling and classroom context. In addition, candidates utilize a matrix to input their collected data on each of their pupils (see Table C.5).

After collecting data, students prepare a response paper (approximately five to seven pages) whereby they examine their data for patterns and (re)examine those emergent patterns by drawing on varied course readings and class discussions. They holistically describe their class with respect to the characteristics listed below (which are initial assumptions). They are also encouraged to describe what their students can do well as well as what they are still learning to do. It is from this data collection process that candidates begin Backwards Planning (Wiggins & McTighe, 2005) for instructional units that are intended to draw on students' backgrounds, needs, and interests. The planning rationale must explicitly justify

Table C.5. Classroom Portrait

Pupil's Name	Data Collection Matrix							
	Home Lang.	**Lang. Dev.**	**Social Dev.**	**Academic Dev.**	**Socio economic and cultural context**	**Attitudes toward learning the content area, e.g., reading, writing, and speaking?**	**Individ-ualized Educational Plan (IEP) or other special needs**	**Other facts, observations, insights, special skills, talents, etc.**
Ex: John Doe	English	Speaks English at home and uses informal English and slang. The student usually speaks more Standard English when I or my CT speaks to him. He uses more informal language with the guys sitting in proximity to him.	Quiet student; spends a lot of time reading car magazines in class; interacts with a group of guys sitting close by, but never disruptive. He tends to hang out with one other guy a lot during lunch or other passing periods. He generally has on ear phones and listens to music.	Average student; earns mostly B grades; scores well on standardized tests according to teacher data. I sense that this student does enough to get by, but I feel that he has great potential to earn an A in my course.	Live in the local community; is from a working-class home; parents attended college and father graduated; parents both work outside the home. He always is well-dressed and appears to pay a good deal of attention to fashion and latest designers and trends.	Enjoys reading magazines; doesn't like writing very much; doesn't typically speak much in class. I would like to see him develop his academic voice more.	No IEP on file	Drives; enjoys sports; works after school a few days per week; wants to go to college; loves to draw and keeps a sketch book—sometimes draws when should be focused on class work.

how the data from the classroom portrait shapes the preliminary instructional purpose, goals, objectives, activities, and assessments. A candidate who teaches Spanish shares the following:

> I utilized information from my classroom portrait to influence decisions I've made in the classroom by looking at where my students come from, their first languages and learning styles. I have some native speakers of Spanish and therefore tried to challenge them, while still keeping the lesson at the appropriate level for the entire class. I also have considered students that come from various cultures speak the languages like Urdu, Arabic and Punjabi and have encouraged them to share what they already know about languages and help the class make connections across them . . . I commonly ask the students what the word for something is in their own language to see if there are any common threads from the content to their own roots. . . . I have some learners with special needs who need extra supports in order to engage with content (graphic organizers, sentence starters, study guides, and reading strategies).

Finally, candidates were required to teach demonstration lessons during our class time throughout the semester. They were placed in interdisciplinary teaching teams throughout the spring semester to develop and create five mini-units (each with two to three lessons) related to *each* thematic area we studied throughout the semester (disability, heterosexism, sexism, linguicism, and the intersection between race and class). After each demonstration lesson, the class would dialogue about the content and skills emphasized in the lesson and provide feedback about the quality of the lesson and the overall social justice purpose and goals. By mandating that candidates develop these lessons, I intentionally chose to disrupt the tendency to treat candidates as technicians who use other people's curriculum in the classroom to teach goals constructed by outsiders—bureaucrats and textbook publishers. Instead, collaboration and planning fostered a culture of risk-taking, agency, and innovation, skills and insight that I desire to have candidates replicate in their field placements. In fact, Table C.6 reflects lessons that were created by teams that were also taught in candidates' field placements.

A Final Note

The accountability movement has constructed and produced failing schools, failing teachers, and failing teacher-education programs. System-wide the standardization discourse has reified dysfunction, reinforcing oppressive power relation within and between schools; all of this continues to have detrimental effects on equitable learning outcomes for children of poor and minority communities and the teachers who are committed to serving them. As a result, educational reform has become deformed (Pinar, 2012) with the increased focus on global economic growth, competition, and discourses focused on accountability. Children are more

Table C.6. Sample Lessons Created by Interdisciplinary Teams

Discipline and Key Concepts	Content Purpose	Multicultural Objective	Assessment
Social Science Key concepts: (race, class, and discrimination)	To evaluate the links of race and class and how they have shaped the lives of people, focusing on the history of Japanese Americans during World War II and the legislation that led to this group's internment.	Students will be able to (SWBAT) recognize the challenges of the Japanese during WWII by discussing how different individuals experienced the implications of the legislation. See stories from www.tellingstories.org/internment/muratsu/index.html	SWBAT write reflective essays and diary entries.
Biology Key concepts: (race, class, food deserts, plants)	The purpose of this lesson is to connect ideas from a biology course (vascular and nonvascular plants) to issues of race and class. Students are learning what plants need to survive and what people need to survive; this is then contrasted with the obstacles of trying to obtain those necessities such as food. People of lower socioeconomic status, who are often also people of color, do not always have access to the nutritious food that people of upper socioeconomic status often take for granted. With this lesson students discuss the causes and effects of this disparity as well as some possible solutions.	SWBAT examine the characteristics of food deserts, both across the United States and in our local area by expressing their opinions during an activity and after viewing video clips ("Let's Move" with Michelle Obama and "Grow Haus—Urban Gardens" (www.video.pbs.org/video/2151097956).	SWBAT write a letter to advocate for solutions to food deserts.
U.S. History Key concepts: (Birmingham campaign, nonviolent direct action, Project C, desegregation, Civil Rights Act of 1964, violent action)	This lesson attempts to meet the Single Group Studies model of multicultural education by Sleeter and Grant because this lesson promotes structural equality for, and immediate recognition of, the identified group under study. Building on several lessons about the civil rights movement, ranging from philosophical similarities and differences between Dr. Martin Luther King Jr. and Malcolm X, this lesson focuses on the Birmingham campaign and the strategies used to fight for civil rights.	SWBAT recognize the different strategies of non-violence and violence used throughout the Birmingham campaign by reviewing the strategies we have learned in previous lessons. SWBAT identify the people involved and the reasons why the campaign for civil rights was successful. Additionally students will explore the sacrifices and the consequences people faced to achieve civil rights. Students will recognize that the civil rights movement was an extraordinary movement done by ordinary people.	Opinion pieces on whether or not they believe the Birmingham campaign was a success or failure? Extended focus: Think about issues that people face today. Is the fight against racism over in this country, or is there more to do? Explain using opinions and examples.

Discipline and Key Concepts	Content Purpose	Multicultural Objective	Assessment
Economics Key concepts: (sexism, demand, profit, incentive, consumer, capitalism, market, corporation)	The purpose of this lesson is for students to apply what they have learned across units in an Economics course to perform an analysis of market capitalism and its role in promoting sexism in society. This lesson is inspired by both the curriculum transformation approach and the social action approach (Banks, 2007) in that the lesson rethinks the curriculum from a gendered, equity perspective and calls on students to make real choices and assert real opinions about their role as consumers in a market system.	SWBAT formulate an opinion of how sexism is related to economics, specifically to the concepts of supply, demand, profit, and consumer-producers relationships, by reading an article and in a jigsaw activity and completing a graphic organizer as a group. Students will read: "Rise in Sexualized Images of Women," www.newswise.com/articles/views/579506; :Forever 21 Selling Allergic to Algebra T-Shirt," www.washingtonpost.com/blogs/blogpost/post/forever-21-selling-allergic-to-algebra-shirt/2011/09/12/gIQAbPqDNK_blog.html; JC Penny, www.washingtonpost.com/blogs/blogspot/post/jcpenny-promotes-im-too-pretty-to-do-homework-shirt/2011/08/31/gIQAxFD4rJ_blog.html	Gallery walk and dialogue about what students learned from readings and images related to sexism and economics
English Key concepts: (linguicism, prejudice, and civil rights)	Students will analyze authentic documents (news articles) concerning Arizona's Department of Education decision to remove "heavily accented" teachers and teachers with speech that deviates from Standard English. This controversial and provocative topic will provide students with the opportunity to examine the issue, take a side, and support their choice in a well-organized and thorough persuasive essay that uses academic language. This is a lesson in the name of social justice because it challenges students and provides them with the tools needed to identify, deconstruct, and refute linguicism.	SWBAT deconstruct and combat linguicism by analyzing and discussing issues surround the recent decisions taken by the Arizona department of education. News articles include "Arizona Seeks to Reassign Heavily Accented Teachers"; "Heavily Accented Teachers Removed from Arizona Classrooms"; "Arizona Teacher Accent Scrutiny Halted"; "Workshop Helps Foreign-born Educators Lose Accent"; "In Arizona, Complaints That an Accent Can Hinder a Teacher's Career."	Students will write a persuasive education in which they take a position for or against the decision made by the Arizona department of education. A clear position stated in the thesis, sufficient evidence to support their claims (from the articles) and good organization using the Jane Schaffer writing strategy. Students will need to compose at least five paragraphs with an introduction, body, and conclusion.

Discipline and Key Concepts	Content Purpose	Multicultural Objective	Assessment
Biology Key concepts: (linguicism, access to healthcare advocacy, civil rights)	This lesson is part of a larger unit on the human body. As a whole, this unit aims to build within students a desire and knowledge to advocate for their health. In addition to learning about the human body and its systems, we will discuss the intersections of race, ethnicity, sex, and language with disease prevalence and access to healthcare. The specific purpose of this lesson is for students to analyze how language (not English) can be a barrier for access to healthcare, thereby increasing disease prevalence and medical complications among non-English speakers. This lesson is inspired by both the curriculum transformation approach and the social action approach (Banks, 2007) because this lesson informs, calls attention to, and promotes advocacy among my students related to inequitable issues in health and healthcare.	SWBAT assess how linguicism affects health and access to healthcare by citing at least three pieces of evidence in their reflective journals. SWBAT identify at least three ways that they can fight/stop linguicism and ways in which organizations are fighting/stopping linguicism as it relates to healthcare access and other aspects of a quality life. Video: Would you Rather . . . (Language Discrimination) at www.youtube.com/watch?v=WV9UmXuYtnI&feature=youtube and 20/20 Linguistic Profiling at www.youtube.com/watch?v=3KCL97s11Jg	Reflection journal, anticipation guides with evidence after readings

than test scores and teachers are more than technicians in the global education marketplace. Thus, I am cognizant of this current historic moment mandates that I, as a teacher-educator, take a more conscientious approach to the preparation of emergent teacher-candidates. As such, I significantly modified my instructional approach within my year-long multicultural education course in a teacher preparation program.

After reviewing of end-of-the-year portfolios that contained varied signature assignments and artifacts from the field placement, I maintain that the training and assessment of emergent teachers must be shaped by critical consciousness borne of embodied praxis—the subjective that shapes learning objectives. This means that the educational journey of the emergent teachers shapes the teaching and learning context, setting the purpose for the learning activity their students will experience. This is a process that needs to be cultivated through risk-taking, practice, dialogue, collaboration, and reflection. For instance, candidates' rationalizations for the lessons that they constructed were often grounded in candidates' ideas about the population they were teaching, which included possessing knowledge about issues shaping their students' day-to-day lives (immigration, racism, poverty, and bullying, to name a few). Second, the portfolio assessment also enabled candidates to use a common language to reflect on their praxis and to engage in dialogue about their efforts to actualize transformative teaching that counters a focus on reductionist and scripted learning in their field placements. I believe that creating space for interdisciplinary planning and discussion related to varied subjectivities challenged candidates to envision alternative possibilities for varied types of disciplinary opportunities for more culturally relevant, content-based instructional topics.

Candidates were guided through a process whereby they were expected to engage in a "discourse of democratic public life" and critique (Giroux, 1997, p. 241) in a classroom space intended to examine hegemonic power relations in American society. Overall, these assessments served as one teacher-educator's attempt to intensively examine the teaching landscape while aiming to reconceptualize an approach to learning that transcended her college's focus on standardized assessment measures. And because so much of the teacher preparation program focuses on the PACT that reifies skills-based approaches to content learning and a contrived production of teaching competency, it became essential to engage candidates in a process of assessment that forced them to draw on their daily practices in their field placements, rather than a videotaped "performance" of a teaching segment. Thus, mandating that candidates collect artifacts from their planning and fieldwork over an entire semester while using those materials to speak about their own development provided a rich and powerful way to gauge emergent teaching competency, particularly in the effort to actualize the TPEs.

In addition, I must also reiterate that emergent teacher competence means more than meeting federal or state definitions of teacher-readiness. Competent teachers disrupt a focus on reductionist learning goals or what Kohn (2011) cites as a "pedagogy of poverty." Readiness from my standpoint means that a teacher demonstrates cultural competence, which includes an open mind, a commitment toward growth and risk-taking, and the ability to reflect on practice. An excerpt from a science candidate shows her views about her growth:

> After completing the portfolio last semester, I did not fully understand all of the different ways in which a curriculum can truly be transformed to be differentiated, equitable, civically engaged and social-justice oriented. Yet, I feel that I have begun to understand and practice this type of pedagogy. . . . I know that teaching takes heart, voice, and allies. I know that I still have a long way to go, but that one aspect of me that I am most proud of is my willingness to have an open mind, to take criticism and to constantly reflect and refine my practice. . . . As we moved through this semester, reading and discussing about other real intersections that our students are affected by (heteronormativity, disabilism, sexism, race, class, and linguicism), I began to see how I could dig deeper and create a curriculum that was truly reflective of these issues. Writing these essays gave me the roots, the backbone and the courage to stand-up and confront my students when they were bullying one another. Furthermore, they gave me the ideas and understandings required to empower myself in finding the connections between my content area and the real skills and understanding my students need to have. In this portfolio, I have included many different artifacts. There are also many

> more that have not been included; they are unplanned conversations with my students . . . the before-school interventions with students who are technically not mine, but actually are if on was to view a school as a community like I do. As I have learned, science will not lie to you; it will say that evidence must be tangible and quantifiable, but this is not true. Evidence of teaching, of a social justice teacher should be seen in their lesson plans, but lesson plans and artifacts are just a part of it. Social justice teaching is the way in which we communicate with our students, the teachable moments that we seize when our students present them, and the offerings from our soul that guide our students patiently in the direction *of their* choosing . . . I can honestly say that I am a different teacher because of our discussions surrounding these topics.

Finally, teaching candidates must be empowered to make connections between the classroom and schooling context and the intersections between their disciplines and the sociopolitical context of the United States. For example, though a sampling of candidates' lessons are presented in Table C.6, the majority of the lesson plans submitted by candidates in their portfolios were lesson attempts that broached the intersections of race, class, gender, etc. In addition, each of the aforementioned scaffolds and assessments were intended as steps toward more authentic assessment that required candidates to link theory and practice in real-world contexts. Standardized testing and grading reduces the quality and quantity of what is taught and learned in schools (McNeil, 2009). Thus it is imperative that teacher-candidates, like their students, are exposed to an education that "teaches them to think critically about knowledge and social institutions and locate their own history and cultural identity within broader contexts" (Lipman, 2009, p. 373). And teacher educators need to reconceptualize learning spaces that offer linkages between theory and practice, shaped by a pedagogy of hope and a belief in the potential of emergent educators to transform learning spaces as sites for democratic praxis.

Endnotes

1 The National Council for Accreditation of Teacher Education (NCATE) is the profession's mechanism to help establish a performance-based system of accreditation With P–12 schools playing a more significant role in designing preparation programs, selecting candidates, assessing candidate performance and progress, and placing them in clinical experiences is a major focus of the reform efforts taking place within NCATE. See http://www.ncate.org/

2 *Brown v. Board of Education*, 347 U.S. 483 (1954), was a landmark U.S. Supreme Court case in which the court declared state laws establishing separate public schools for black and white students unconstitutional.

3 *Lau v. Nichols*, 414 U.S. 563 (1974), was a civil rights case that was brought by Chinese American students living in San Francisco, California, who had limited English proficiency. The students claimed that they were not receiving special help in school due to their inability to speak English, help that they argued they were entitled to under Title VI of the Civil Rights Act of 1964 because of its ban on educational discrimination on the basis of national origin.

4 NCLB supports standards-based education reform based on the premise that setting high standards and establishing measurable goals can improve individual outcomes in education.

5 See http://www.pisa.oecd.org/pages/0,3417,en_32252351_32235907_1_1_1_1_1,00.html

6 This goal was to be reached by the 2005–2006 school year.

7 California created a system for HQT school compliance in 2006 called Compliance, Monitoring, Intervention, and Sanctions (CMIS), which was authorized and funded by the California Legislature. See http://www.cde.ca.gov/nclb/sr/tq/

8 See http://www.ctc.ca.gov/educator-prep/standards/EPPS-Handbook-CTEL.pdf

9 PACT was initiated by a consortium of schools of education at all the Universities of California (or UCs), Mills College and Stanford University, joined early on by three California State Universities or CSUs.

10 On average, students were enrolled in 18–24 units per semester, which often comprised program pre/co-requisites. In addition, 7 units of students' course load is their field placement teaching for 2–3 class period per day.

11 During the 2011–2012 academic year, 14 students (3 males; 11 females) were enrolled in the cohort. The demographics were as follows: six white; three Asian; five Latino. The course met one evening per week (3 hours) during the fall and spring semesters.

12 Students typically speak no more than 20 minutes.

13 Students write a 5–7 page reflection paper and speak to how their artifacts reflect varied TPEs. This paper serves as a springboard for follow-up dialogue about areas that are strong, and also those areas that demand further consideration and strengthening.

14 This reflection is submitted at the end of each semester.

15 Seek information from National and State Standards for Teaching in the Content Area; from the mission and objectives of the National Organization specific to your discipline (National Council of Teachers of English; National Council for the Social Studies; National Science Teachers Association; National Council of Teachers of Math; and the American Council of the Teaching of Foreign Languages); from the National or the California Association for Bilingual Education (NABE/CABE); and from the National Association of Multicultural Education (NAME). These types of professional organization have mission statements and position papers on varied issues related to teaching and learning.

References

Banks, J. A. (2008). *Introduction to multicultural education* (4th ed.). Boston: Allyn & Bacon.

Berlak, A. (2008, June). Coming to a credential program near you—National exit exams for teachers. *Rethinking Schools.* Retrieved June 10, 2012, from http://www.rethinkingschools.org/restrict.asp?path=archive/24_04/24_04_exams.shtml

Berta-Avila, M., & William-White, L. (2010, October). A conscious agenda for cultivating future teachers' equity and social justice paradigms. *Teacher Education and Practice, 23*(4), 1–37.

Common Core State Standards Initiative. (n.d.). *About the Standards.* Retrieved April 22, 2012, from http://www.corestandards.org

Crowe, E. (2010). Measuring what matters: A stronger accountability model for teacher education. Center for Education Progress. Retrieved June 12, 2012, from http://www.americanprogress.org/issues/2010/07/pdf/teacher_accountability.pdf

From teasing to torment: School climate in America—a survey on students and teachers (2005, October 11). Gay, Lesbian, Straight Education Network. Retrieved June 12, 2012 from http://www.glsen.org/binarydata/GLSEN_ATTACHMENTS/file/499-1

Freire, P. (1971). *Pedagogy of the Oppressed.* New York: Seabury.

Fullan, M. G., & Stiegelbauer, S. (1991). *The new meaning of educational change.* New York: Teachers College Press.

Giroux, H. A. (1997). *Pedagogy and the politics of hope: Theory, culture, and schooling.* Boulder: Westview Press.

Hawkins, S. (2010, December). Education vs. incarceration. The American Prospect. Retrieved June 7, 2012, from http://prospect.org/article/education-vs-incarceration

Highly Qualified Teachers in the United States: Where are we now? (2008, May). National Comprehensive Center for Teacher Quality. Retrieved May 28, 2012, from http://www.tqsource.org/publications/HQTinUS.pdf

Kohn, A. (2000, September). Standardized testing and its victims. *Education Week*. Retrieved August 8, 2012, from http://www.alfiekohn.org/teaching/edweek/staiv.htm

Kohn, A. (2011, April). Poor teaching for poor children . . . In the name of reform. *Education Week*. Retrieved August 8, 2012, from http://www.alfiekohn.org/teaching/edweek/poor.htm

Ladson-Billings, G. (1998). Just what is critical race theory and what is it doing in a "nice" field like education. *International Journal of Qualitative Studies in Education, 11*, 7–24.

Lipman, P. (2009). Beyond accountability: Towards schools that create new people for a new way of life. In A. Darder, M. P. Baltodano, & R. D. Torres (Eds.), *The Critical Pedagogy Reader* (2nd ed., pp. 364–383). New York: Routledge.

McNeil, L. (2009). Standardization, defensive teaching, and the problem of control. In A. Darder, M. P. Baltodano, & R. D. Torres (Eds.), *The Critical Pedagogy Reader* (2nd ed., pp. 384–396). New York: Routledge.

Measuring Child Poverty: New league tables of child poverty in the world's richest countries. (2012, May). UNICEF. Retrieved June 12, 2012, from http://www.unicef-irc.org/publications/pdf/rc10_eng.pdf

Meeting the Highly Qualified Teacher Challenge: The secretary's second annual report on teacher quality. (2003, June). U.S. Department of Education Office of Postsecondary Education. Retrieved May 28, 2012, from http://www2.ed.gov/about/reports/annual/teachprep/2003title-ii-report.pdf

Misplaced Priorities: Over incarcerate, under education. (2011, May). National Association for the Advancement of Colored People's Smart and Safe Campaign. Retrieved June 7, 2012, from http://naacp.3cdn.net/ecea56adeef3d84a28_azsm639wz.pdf

Moll, L. C., Amanti, C., Neff, D., & Gonzalez, N. (1992). Funds of knowledge for teaching: Using a qualitative approach to connect homes and classrooms. *Theory into Practice*, 31(2), 132–141.

Moule, J. (2012). *Cultural competence: A primer for educators* (2nd ed.). Belmont, CA: Wadsworth.

Performance Assessment for California Teachers (PACT). Retrieved June 10, 2012, from http://www.pacttpa.org/_main/hub.php?pageName=Supporting_Documents_for_Candidates

Pinar, W. F. (1994). *Autobiography, politics, and sexuality: Essays in curriculum theory (1972–1992)*. New York: Peter Lang.

Pinar, W. F. (2012). *What is curriculum theory* (2nd ed.). New York: Routledge.

Ronfeldt, M., Loeb, S., & Wyckoff, J. (2012). Paper presented at the Center for Longitudinal Data in Education Research. Retrieved June 14, 2012, from http://blogs.edweek.org/edweek/teacherbeat/TchTrnStAch%20AERJ%20R%26R%20not%20blind.pdf

Sleeter, C. E., & Grant, C. A. (2003). *Making choices for multicultural education: Five approaches to race, class, and gender.* New York: Wiley.

State Highly Qualified Teacher (HQT) Plans. (2012). National Comprehensive Center for Teacher Quality (2012). Retrieved May 28, 2012, from http://www.tqsource.org/topics/HQT.php

Strengthening Our Military Families: Meeting America's Commitment. (2011). Retrieved June 14, 2012, from http://www.defense.gov/home/features/2011/0111_initiative/strengthening_our_military_january_2011.pdf

Teel, K. M., & Obidah, J. E. (Eds.). (2008). *Building racial and cultural competence in the classroom: Strategies from urban educators.* New York: Teachers College Press.

Watt, C. D., Proctor, B. D., & Smith, J. C. (2011, September). Income, poverty, and health coverage in the United States: 2010. Retrieved June 14, 2012, from http://www.census.gov/prod/2011pubs/p60-239.pdf

Wiggins G., & McTighe, J. (2005). *Understanding by design* (2nd ed.). (CITY), NJ: Pearson Education.

William-White, L., Muccular, D., Muccular, G., & Brown, A. (2013). *Critical consciousness in curricular research: Evidence from the field.* New York: Peter Lang.

Author Biographies

Lawrence Baines is professor and chair of instructional leadership at The University of Oklahoma. His latest books are *The Teachers We Need* (Rowman & Littlefield, 2010) and *Going Bohemian: Teaching Writing Like You Mean It* (International Reading Association, 2010). Check out www.lawrencebaines.com.

Brian R. Beabout is an assistant professor of education at the University of New Orleans. He is the associate editor for the journal *Complicity: An International Journal of Complexity and Education* and is on the editorial review board for the *School Community Journal.* His work has been published in the *Journal of Thought, The Journal of Educational Change, The Journal of Education for Students Placed At-Risk, Multicultural Education,* and *The International Journal of Educational Reform.* He was a founding board member of the Morris Jeff Community School in New Orleans and can be reached at bbeabout@uno.edu or @bbeabout.

David L. Bolton is an associate professor at West Chester University in Pennsylvania. He received his PhD in research and testing from Florida State University and his M.S. in research and statistical methodology from Andrews University. His undergraduate degree is in Theology from Seminar Marienhoehe in Darmstadt, Germany. Dr. Bolton teaches educational assessment, as well as research and statistics. His areas of scholarly activity have focused upon the use of assessment and technology in education.

Joe Bower is a teacher in Red Deer, Alberta, Canada. He spent 10 years teaching grade 8 language arts and science in a middle school, and he is now a special education teacher in a children's psychiatric assessment unit. Follow his work at www.joebower.org and @joe_bower.

Anthony Cody received his teaching credential at UC Berkeley in 1987 and taught science and math for the following 18 years at a high-needs middle school in Oakland. In 2007, he initiated a district-wide mentoring project in Oakland called TeamScience. He now lives in Mendocino County and authors the popular *Education Week* blog Living in Dialogue, which discusses education policy issues.

Peter DeWitt, EdD, is an elementary principal in upstate New York and blogs at Finding Common Ground for *Education Week*. He has written articles at the state, national, and international levels. His is the author of *Dignity for All: Safeguarding LGBT Students* (Corwin). He can be found at www.petermdewitt.com.

John M. Elmore is currently an associate professor of professional education at West Chester University of Pennsylvania. Dr. Elmore was previously an associate professor and director of graduate studies at Medaille College in Buffalo, New York, from 2000 to 2005. He holds degrees in history (1995) and psychology (1996) from Kansas Wesleyan University along with an MS in Educational Foundations (1997) and PhD in Curriculum and Instruction (2000) from Kansas State University. Recent publications include "War and the Sectarian Mind" in *Educating for Peace in a Time of Permanent War* (Routledge, 2012) and *Teaching Marx: The Socialist Challenge* (Information Age, 2012).

Hadley J. Ferguson is a middle school teacher at Springside Chestnut Hill Academy in Philadelphia. She is a member of the Library of Congress Teaching with Primary Sources (TPS) Mentor Advisory Group, as well as a founding board member of the Edcamp Foundation. She can be found at her blog, Middle School Matrix, and @hadleyjf.

Rhonda Goolsby is an assistant professor in the Instructional Leadership and Academic Curriculum Department in the College of Education at The University of Oklahoma. Previously, Rhonda taught elementary and middle school in Texas and studied at Texas A&M University in College Station, Texas. Her research also includes Dual Coding Theory and struggling readers.

David A. Gorlewski is an assistant professor in the Education Department at D'Youville College in Buffalo, New York, where he teaches courses in curriculum planning, methods, and advanced curricular issues. He has served in public education as a high school English teacher, a staff developer, and a senior level administrator in curriculum and personnel. Most recently, he co-authored with Julie Gorlewski *Making It Real: Case Stories for Secondary Teachers* (Sense, 2012) and co-edited with Bradley Porfilio and Julie Gorlewski *Using Standards and High-Stakes Testing* for *Students: Exploiting Power with Critical Pedagogy* (Peter Lang, 2012).

Julie A. Gorlewski is an assistant professor of secondary education at the State University of New York at New Paltz and taught secondary English for more than 15 years. She is the author of the book *Power, Resistance, and Literacy: Writing for Social Justice* (Information Age, 2011), which was selected for a 2011 Critic's Choice Award by the American Educational Studies Association. In addition, she co-authored with David Gorlewski *Making It Real: Case Stories for Secondary Teachers* (Sense, 2012) and co-edited with Bradley Porfilio and David Gorlewski *Using Standards and High-Stakes Testing* for *Students: Exploiting Power with Critical Pedagogy* (Peter Lang, 2012).

After many years as a writer and editor, **Lisa Guisbond** became interested in education policy as a public school parent with an interest in special education. She was the lead author of FairTest's report "NCLB's Lost Decade for Educational Progress" and co-authored "Failing Our Children: How 'No Child Left Behind' Undermines Quality and Equity in Education." She is vice president of Citizens for Public Schools and is the principal author of the "Campaign for the Education of the Whole Child." Her writing on education and assessment has appeared in a wide range of publications, including *Education Week*, the *Washington Post*, the *Atlanta Journal Constitution*, *USA Today*, and the *New York Times*. She served on the MCAS and Assessment subcommittee of Massachusetts Governor Deval Patrick's Readiness Project.

A former lawyer, **John Hoben** is a recent graduate of the PhD program at Memorial University's Faculty of Education. John's thesis work was funded through a Canada Graduate Scholarship from the Social Sciences and Humanities Research Council of Canada. This research was a qualitative study that sought to describe teacher perceptions of legal rights and their effect on school governance and teaching practice. John's teaching and writing demonstrate broader concerns with schooling and social justice, narrative education and the importance of creating meaningful, collaborative working relationships and spaces where students can bring imagination and personal experience to their academic work.

Alfie Kohn is the author of 12 books on education, parenting, and human behavior, including *Punished by Rewards* (1993), *Beyond Discipline* (1996), *The Schools Our Children Deserve* (1999), *Unconditional Parenting* (2005), *The Homework Myth* (2006), and, most recently, *Feel-Bad Education* (2011). *Time* magazine has described him as "perhaps the country's most outspoken critic of education's fixation on grades [and] test scores." Kohn works with educators and parents across the country and speaks regularly at national conferences. He lives (actually) in the Boston area and (virtually) at www.alfiekohn.org.

Morna McDermott is an associate professor at Towson University, where she teaches various theory and methods courses in the College of Education. Her scholarship and research interests focus on democracy, social justice, and arts-informed inquiry in K–postsecondary educational settings, and working with beginning and experienced educators. Dr. McDermott is one of the administrators for the national organization United Opt Out National in the fight against corporate reform, maintains a regular blog, and has a forthcoming book with Information Age Press titled *The Left Handed Curriculum: Empowering Educators through Creativity*.

Richard Mora is an assistant professor of sociology at Occidental College. His research interests include education, youth cultures, gender, and juvenile justice. He has published in *Gender & Society*, *Journal of Inquiry and Action in Education*, and *Perspectives on Urban Education*, and he has work included in *The Phenomenon of Obama and the Agenda for Education: Can Hope Audaciously Trump Neoliberalism?* (Information Age, 2011) and *Encyclopedia of School Crime and Violence* (Greenwood, 2011). He can be reached at rmora@oxy.edu.

Monty Neill, EdD, is executive director of FairTest: National Center for Fair & Open Testing and chair of the national Forum on Educational Accountability. He has initiated national and state coalitions of education, civil rights, religious, disability, parent and other organizations to work toward fundamental change in student assessment and school accountability. His many publications address problems with testing, benefits of high-quality assessment, and resistance to high-stakes testing. FairTest's website is http://www.fairtest.org.

Fernando F. Padró is senior lecturer (quality) at the University of Southern Queensland. He is also editor of the American Society for Quality Education Division *Quality Approaches in Higher Education*, serves as a NASPA Faculty Fellow, and is a member of INQAAHE's Best Practices Committee. His major research interest is quality assurance policy and practice in higher education and he has served as an examiner for different government agencies and accrediting bodies in the United States. His publications in quality assurance are based on his research and experiences as an administrator as well as member of different academic faculties over the years.

Andre M. Perry, PhD, is the associate director for educational initiatives for Loyola Institute for Quality and Equity in Education. Prior to, he helped manage four charter schools in post-Katrina New Orleans. In 2011, UNO Press released his work of fiction, *The Garden Path: The Miseducation of the City*. Perry's views, opinions, and educational leadership have been featured on National Public Radio and CNN. In 2010, Perry served on Mayor-Elect Mitch Landrieu's Transition Team as the co-chair of the Education Taskforce. He also serves as Team Lead for Orleans PLACE MATTERS, a national initiative out of the Joint Center for Political and Economic Studies.

Brian Rhode, MEd, has spent the past 12 years working in elementary education. He has taught first, second, and fifth grades so far in his career. During the summer of 2009 Brian spent three weeks reflecting on, discussing, practicing, and researching effective methods for teaching writing as part of the Capital District Writing Project's 2009 Summer Institute. Brian continues to work with the Capital District Writing Project both as a consultant and presenter at various events including their 2011 conference "Teaching to the Core." Brian lives near Albany, New York, with his wife and two children. He can be contacted through brianrhode77@gmail.com.

Bob Schaeffer has served as public education director of FairTest, the National Center for Fair & Open Testing, since its founding in 1985. He also is a member of the organization's Board of Directors. Prior to the creation of FairTest, Mr. Schaeffer was editorial writer at the NBC-TV affiliate in Boston and research director of the Massachusetts Legislature's Joint Committee on Human Services and Elderly Affairs. He also worked for several years as a research associate at the Education Research Center of the Massachusetts Institute of Technology (MIT), where he was both an undergraduate and graduate student. Bob Schaeffer is the author of *Standardized Tests and Teacher Competence* (School Voices, Fall, 1996) and coauthor of *Standing Up to the SAT* (ARCO/Simon & Schuster, 1989).

P. L. Thomas, associate professor of education (Furman University, Greenville, South Carolina), taught high school English in rural South Carolina before moving to teacher education. He is currently a column editor for *English Journal* (National Council of Teachers of English) and series editor for *Critical Literacy Teaching Series: Challenging Authors and Genres* (Sense Publishers). Follow his work at http://wrestlingwithwriting.blogspot.com/ and @plthomasEdD.

Jim Webber is an assistant professor of English at the University of Nevada, Reno. Previously, Jim taught high school English in Vermont and studied composition and rhetoric at the University of New Hampshire. His writing has appeared in *Kappan*, *Rhetoric Society Quarterly*, and *JAC*.

Lisa William-White is an associate professor of education at California State University Sacramento. Samples of her work appear in *Tedious Journeys: Auto-*

ethnography by Women of Color in Academe (Peter Lang); *Hip-Hop(e): The Cultural Practice and Critical Pedagogy of International Hip-Hop Inquiry* (Peter Lang); *Choosing Democracy: A Practical Guide to Multicultural Education*, 3rd edition (Merrill/Prentice Hall); *Teacher Education Quarterly; Qualitative Inquiry; Qualitative Research in Education; International Journal of Qualitative Studies in Education; International Review of Qualitative Research and Cultural Studies* < = > *Critical Methodologies.* She situates her work in the areas of critical qualitative research, critical theory, and multicultural and urban education.

Maja Wilson is the author of *Rethinking Rubrics in Writing Assessment* (Heinemann, 2006). She taught high school English, adult basic education, and alternative high school and middle school in Michigan's public schools for 10 years. She is finishing her doctorate in English at the University of New Hampshire and is an instructor in the literacy program at the University of Maine's College of Education and Human Development. Her writing has appeared in various educational journals.

Studies in the Postmodern Theory of Education

General Editor
Shirley R. Steinberg

Counterpoints publishes the most compelling and imaginative books being written in education today. Grounded on the theoretical advances in criticalism, feminism, and postmodernism in the last two decades of the twentieth century, Counterpoints engages the meaning of these innovations in various forms of educational expression. Committed to the proposition that theoretical literature should be accessible to a variety of audiences, the series insists that its authors avoid esoteric and jargonistic languages that transform educational scholarship into an elite discourse for the initiated. Scholarly work matters only to the degree it affects consciousness and practice at multiple sites. Counterpoints' editorial policy is based on these principles and the ability of scholars to break new ground, to open new conversations, to go where educators have never gone before.

For additional information about this series or for the submission of manuscripts, please contact:

Shirley R. Steinberg
c/o Peter Lang Publishing, Inc.
29 Broadway, 18th floor
New York, New York 10006

To order other books in this series, please contact our Customer Service Department:

(800) 770-LANG (within the U.S.)
(212) 647-7706 (outside the U.S.)
(212) 647-7707 FAX

Or browse online by series:

www.peterlang.com